Florida Algebra I End-of-Course (EOC)

Study Guide 2024-2025

Master the Algebra I End-of-Course Exam with Detailed Content Review, Test-Taking Strategies, and Two Full-Length Practice Tests to Ace the Florida EOC Math Test

Test Treasure Publication

Test Treasure
PUBLICATION

COPYRIGHT

Trademarks

All trademarks, service marks, and trade names used within this website and Test Treasure Publication's products are proprietary to Test Treasure Publication or other respective owners that have granted Test Treasure Publication the right and license to use such intellectual property.

Disclaimer

While every effort has been made to ensure the accuracy and completeness of the information contained in our products, Test Treasure Publication assumes no responsibility for errors, omissions, or contradictory interpretation of the subject matter herein. All information is provided "as is" without warranty of any kind.

Governing Law

This website is controlled by Test Treasure Publication from our offices located in the state of California, USA. It can be accessed by most countries around the world. As each country has laws that may differ from those of California, by accessing our website, you agree that the statutes and laws of California, without regard to the conflict of laws and the United Nations Convention on the International Sales of Goods, will apply to all matters relating to the use of this website and the purchase of any products or services through this site.

CONTENTS

INTRODUCTION

Welcome to the **Florida Algebra 1 End-of-Course (EOC) Study Guide 2024-2025**! Whether you are a student preparing for the exam, a teacher looking for comprehensive resources, or a parent helping your child succeed, this guide is designed to support your learning journey.

The **Florida Algebra 1 End-of-Course (EOC) Exam** is a critical milestone for high school students in Florida. It assesses algebraic concepts and problem-solving skills that are essential for higher mathematics. Success on this exam is not just about passing a test—it is about developing a strong mathematical foundation that will benefit students in advanced courses and real-world applications.

This study guide provides:

✔ **Complete Coverage of All Exam Topics** – Each section is structured according to the Florida Algebra 1 EOC standards, ensuring a thorough review of all tested concepts.

✔ **Clear Explanations & Step-by-Step Solutions** – Difficult algebraic concepts are broken down into easy-to-understand explanations, with worked-out examples and step-by-step problem-solving techniques.

✔ **Practice Questions & Full-Length Tests** – With **over 100+ practice questions** and **two full-length mock exams**, students can build confidence and improve test-taking skills.

✔ **Test-Taking Strategies & Study Plans** – This guide includes expert strategies to help you manage time effectively, reduce test anxiety, and approach different types of questions strategically.

Who Should Use This Book?

- **Students** preparing for the Florida Algebra 1 EOC exam

- **Teachers** looking for structured review materials and practice problems

- **Parents** who want to help their children succeed in Algebra 1

- **Tutors & Homeschoolers** who need a complete resource for guided instruction

How to Use This Study Guide?

Read Each Concept Thoroughly – Start by reviewing the detailed explanations for each algebra topic.

Practice, Practice, Practice! – Solve the provided exercises and check your answers with the detailed explanations.

Take the Full-Length Tests – Simulate the actual exam experience with the included **two full-length practice tests**.

Review Weak Areas – Focus on sections where you struggle and revisit explanations and practice problems.

Use the Study Plans – Follow the **customized study schedules** to stay on track and ensure full preparation before the exam.

Final Words

Mathematics is a skill that improves with **practice and patience**. This study guide will not only help you **pass the Florida Algebra 1 EOC exam** but will also build your confidence in algebraic thinking. Stay dedicated, work through the problems, and believe in your ability to succeed!

Brief Overview of the Exam and Its Importance

The **Florida Algebra 1 End-of-Course (EOC) Exam** is a **state-mandated assessment** designed to evaluate students' understanding of key algebraic concepts. This exam is an essential part of the Florida high school curriculum and serves as a **graduation requirement** for students enrolled in Algebra 1.

Success on this exam is not only important for **fulfilling state educational requirements** but also for building a **strong foundation in mathematics**, which is crucial for future coursework in **geometry, advanced algebra, calculus, and STEM-related careers**.

Exam Overview

Feature	Details
Administered By	Florida Department of Education (FLDOE)
Exam Format	Computer-Based Test (CBT)
Number of Questions	Approximately **60–66** multiple-choice and open-response questions
Time Limit	**160 minutes** (Divided into two 80-minute sessions)
Question Types	Multiple-Choice, Multi-Select, Grid-In (Open-Response), Drag-and-Drop, Fill-in-the-Blank
Passing Score	Scale score of **497 or above** on a **425–575 scale**
Scoring Levels	Level 1: 425–486 (Inadequate)
	Level 2: 487–496 (Below Satisfactory)
	Level 3: 497–517 (Satisfactory – Passing Score)
	Level 4: 518–531 (Proficient)
	Level 5: 532–575 (Mastery)
Testing Window	Conducted **multiple times per year** (Spring, Summer, and Fall)
Retake Policy	Students who do not pass can **retake the test** in subsequent testing windows
Calculator Policy	A **scientific calculator** is allowed for **Session 2 only**
Exam Platform	Administered on **Florida's TestNav system**

Importance of the Florida Algebra 1 EOC Exam

1. Graduation Requirement

The **Florida Algebra 1 EOC exam is a requirement for high school graduation**. Students must pass this exam to earn their **high school diploma** unless they meet the criteria for an alternate assessment.

2. College and Career Readiness

Algebra 1 is **the foundation of higher-level mathematics**. Many careers in **engineering, finance, medicine, computer science, and technology** require strong algebra skills. Passing this exam demonstrates proficiency in fundamental math concepts.

3. Course Progression

A **strong performance** on the Algebra 1 EOC exam is essential for progressing to higher-level courses, such as **Geometry, Algebra 2, Pre-Calculus, and Calculus**. Students who struggle may need additional remediation before advancing.

4. Performance-Based Placement

Some Florida schools use **EOC exam scores** for **class placement** in advanced math courses. A high score may qualify students for **accelerated or honors-level math classes**.

5. College Entrance Exams (SAT/ACT Readiness)

Many questions on the **SAT, ACT, and other standardized tests** are based on **Algebra 1 concepts**. Performing well on this exam helps students prepare for these **college entrance exams**.

6. Alternative Assessments for Failing Students

Students who do not pass the Algebra 1 EOC exam can meet the **graduation requirement** by achieving a minimum required score on:

- **PSAT/NMSQT (Score of 430 or higher on Math section)**

- **SAT Math (Score of 420 or higher)**

- **ACT Math (Score of 16 or higher)**

Exam Structure Breakdown

The **Florida Algebra 1 EOC exam** covers **seven core areas**, which are tested through various question formats.

Test Section	Topics Covered	Weight (%)
Algebra Foundations	Properties of Real Numbers, Order of Operations, Simplifying Expressions, Solving Equations & Inequalities, Absolute Values	10–15%
Linear Equations & Functions	Graphing Equations, Writing Linear Equations, Solving Systems of Equations, Modeling with Linear Functions	30–35%
Quadratic Equations & Functions	Graphing Quadratic Functions, Solving Quadratics, Quadratic Inequalities, Applications	25–30%
Exponential Functions	Exponential Growth & Decay, Properties of Exponents, Solving Exponential Equations	10–15%
Polynomials & Factoring	Adding, Subtracting, Multiplying, Factoring, Solving Polynomial Equations	5–10%
Radical & Rational Expressions	Simplifying, Rationalizing, Solving Radical & Rational Equations	5–10%
Data Analysis & Probability	Interpreting Graphs & Tables, Mean, Median, Mode, Probability, Scatter Plots	5–10%

Final Thoughts

The **Florida Algebra 1 EOC Exam** is a **crucial step in a student's academic journey**. A strong performance on this exam demonstrates **proficiency in algebraic concepts**, **builds confidence**, and **opens doors for future success in higher mathematics and career fields**.

 Use this study guide to thoroughly review all topics, practice with real exam-style questions, and implement test-taking strategies.

DETAILED CONTENT REVIEW

This **detailed content review** covers all the **key concepts** tested in the **Florida Algebra 1 End-of-Course (EOC) Exam**. Each section includes **definitions, formulas, step-by-step explanations, examples, and problem-solving techniques** to help students master algebraic concepts efficiently.

Section 1: Algebra Foundations

1.1 Properties of Real Numbers

Real numbers follow certain rules that make calculations predictable.

Property	Definition	Example
Commutative	Order of numbers does not affect the sum or product.	$a + b = b + a, ab = ba$
Associative	Grouping of numbers does not affect the sum or product.	$(a + b) + c = a + (b + c)$
Distributive	Multiplication distributes over addition.	$a(b + c) = ab + ac$
Identity	Adding 0 or multiplying by 1 does not change a number.	$a + 0 = a, a \times 1 = a$
Inverse	A number plus its opposite equals zero.	$a + (-a) = 0$

1.2 Order of Operations (PEMDAS)

PEMDAS Rule: Parentheses → Exponents → Multiplication/Division → Addition/Subtraction

Example:

$8+(3\times2)^2\div6$

Step 1: Parentheses → $3\times2=6$
Step 2: Exponents → $6^2=36$
Step 3: Division → $36\div6=6$
Step 4: Addition → $8+6=14$
Answer: 14

1.3 Evaluating Algebraic Expressions

Substituting values into expressions and simplifying.

Example: Evaluate $3a^2-2b+4$ when $a=2, b=5$.

$3(2)^2-2(5)+4=3(4)-10+4=12-10+4=6.$

Answer: 6

1.4 Solving Basic Equations & Inequalities

Solving for x in equations and inequalities.

Example: Solve for x:

$2x-4=10$

Step 1: Add 4 to both sides → $2x=14$
Step 2: Divide by 2 → $x=7$

✔ Inequalities Rule:

When multiplying/dividing by a **negative number**, **flip the inequality sign**!

 Example: Solve $-3x>9$
Step 1: Divide by -3 → $x<-3$ (**flip sign!**)

1.5 Understanding Absolute Values

Absolute value measures distance from zero.

 Example: Solve $|x-4|=6$
Two possible cases:

   1. $x-4=6$ → $x=10$

   2. $x-4=-6$ → $x=-2$
     ✔ **Answer: x=10,−2**

Section 2: Linear Equations & Functions

2.1 Understanding Functions (Definition & Notation)

A function has **one unique output for each input**.

Example: Identify the function:

✔ f(x)=3x+2 (**Function**)

✘ y2=x (**Not a function — one input may have multiple outputs**)

2.2 Graphing Linear Equations (Slope & Intercepts)

Slope Formula:

m=y2−y1/x2−x1

Example: Find slope between (2,4) and (5,10)

m=10−4/5−2=6/3=2

✔ **Answer: Slope = 2**

2.3 Solving Systems of Equations

Methods:

- **Graphing**: Find intersection

- **Substitution**: Solve for one variable and substitute

- **Elimination**: Add/subtract equations

Example (Elimination Method)

3x+2y=10, 2x−2y=4

Adding both: 5x=14 → x=14/5

Section 3: Quadratic Equations & Functions

Standard Form:

$ax^2+bx+c=0$

Methods to Solve:

Factoring

Quadratic Formula:

$x=-b\pm\sqrt{b^2-4ac}/2a$

Completing the Square

Example (Factoring)

$x^2-5x+6=0$

Factor: $(x-2)(x-3)=0$

Solutions: $x=2,3$

Section 4: Exponential Functions

Growth & Decay Formula:

$A=A_0(1+r)^t$

Example: A bacteria population doubles every hour. If there are 100 bacteria initially, how many after 3 hours?

$A=100(2)^3=100(8)=800$

✔ **Answer: 800 bacteria**

Section 5: Polynomials & Factoring

Polynomial Operations:

- Add/subtract like terms

- Multiply using **FOIL method**

- Factor using **GCF, difference of squares, trinomials**

Section 6: Radical & Rational Expressions

Simplifying Square Roots

$\sqrt{50} = \sqrt{25 \times 2} = 5\sqrt{2}$

Solving Rational Equations
Find **LCD**, multiply, solve for x.

Example:

$2/x + 3 = 5$

Multiply by x, solve: x=1

Section 7: Data Analysis & Probability

Mean, Median, Mode, Range

- **Mean**: Average

- **Median**: Middle value

- **Mode**: Most frequent

- **Range**: Difference between highest and lowest

Example: Find the mean of 5,10,15,20

5+10+15+20/4=12.5

Final Words

This **content review** provides a **strong foundation** in Algebra 1 topics. **Practice consistently, use problem-solving techniques, and apply strategies** to master the Florida Algebra 1 EOC Exam!

STUDY SCHEDULES AND PLANNING ADVICE

Preparing for the **Florida Algebra 1 End-of-Course (EOC) Exam** requires a **structured study plan** to cover all essential topics, practice effectively, and build confidence before test day. This section provides **customized study schedules** based on different preparation timelines, along with **expert planning advice** to help you maximize your study sessions.

How to Plan Your Study Schedule?

Step 1: Assess Your Current Knowledge

- Take a **diagnostic test** to identify strong and weak areas.

- Focus more time on topics where you struggle.

- Set realistic goals based on your test date.

Step 2: Choose a Study Schedule Based on Your Timeframe

Below are three different study plans depending on the time you have before the exam:

One-Month Intensive Study Plan (Best for last-minute preparation)
Two-Month Balanced Study Plan (Best for gradual learners)
Three-Month In-Depth Study Plan (Best for comprehensive preparation)

Step 3: Follow a Structured Approach

- Allocate **dedicated study hours** daily or weekly.

- Focus on **understanding concepts** before solving problems.

- Take **regular quizzes** and review mistakes.

- Include **timed practice tests** to simulate the actual exam.

One-Month Intensive Study Plan (4 Weeks)

Best For: Students who have **limited time** and need a **quick, focused review** before the exam.

Week	Focus Topics	Tasks
Week 1	Algebra Foundations & Linear Equations	Review properties of numbers, order of operations, solving equations, inequalities, absolute values. Solve at least **30 practice problems**.
Week 2	Quadratic & Exponential Functions	Learn quadratic functions, solving quadratics, quadratic inequalities, exponential growth/decay. Solve **40+ problems** on these topics.
Week 3	Polynomials, Factoring & Rational Expressions	Master factoring, simplifying polynomials, radical and rational expressions. Solve **at least 50 problems**.
Week 4	Data Analysis & Full-Length Practice Tests	Focus on mean, median, mode, probability, scatter plots. Take **two full-length practice tests**. Review mistakes and refine weak areas.

Daily Time Commitment: 2–3 hours per day
Key Strategy: Focus on **high-yield topics** that appear frequently on the exam.

Two-Month Balanced Study Plan (8 Weeks)

Best For: Students who prefer a **moderate-paced study schedule** while balancing schoolwork.

Week	Focus Topics	Tasks
Weeks 1-2	Algebra Foundations & Linear Functions	Cover order of operations, evaluating expressions, solving equations, functions, graphing linear equations. Complete 50+ practice problems.
Weeks 3-4	Quadratic & Exponential Functions	Learn factoring quadratics, quadratic formula, graphing quadratics, solving exponential equations. Solve 60+ problems.
Weeks 5-6	Polynomials, Rational & Radical Expressions	Master simplifying polynomials, factoring, rationalizing denominators, solving radical equations. Solve 50+ problems.
Weeks 7-8	Data Analysis & Full-Length Practice Tests	Focus on mean, median, mode, probability, scatter plots. Take two full-length practice tests.

Daily Time Commitment: 1–2 hours per day
Key Strategy: Spend **more time reviewing mistakes** and strengthening weak areas.

Three-Month In-Depth Study Plan (12 Weeks)

Best For: Students who want a **thorough review with deeper understanding** and **consistent practice.**

Month	Focus Topics	Tasks
Month 1	Algebra Foundations & Linear Functions	Focus on **real numbers, equations, graphing functions, writing equations**. Solve **75+ problems**.
Month 2	Quadratic, Exponential Functions & Factoring	Master **quadratics, solving equations, exponential functions, factoring polynomials**. Solve **100+ problems**.
Month 3	Advanced Topics & Full-Length Tests	Cover **radical expressions, rational functions, probability, data interpretation**. Take **three full-length practice tests**.

Daily Time Commitment: 1 hour per day / 5 days per week
Key Strategy: Focus on **concept mastery**, **multiple problem-solving techniques**, and **gradual difficulty increase**.

Study Tips & Planning Advice

1. Use a Variety of Study Methods

- **Watch Video Lessons**: Visual learners benefit from online tutorials.

- **Practice with Flashcards**: Memorize formulas and key concepts.

- **Solve Timed Questions**: Build speed and accuracy.

2. Take Regular Breaks

Studying **for hours without breaks** leads to burnout. Use the **Pomodoro Technique**:

- Study for **50 minutes**, take a **10-minute break**.

- Every **3 sessions**, take a **longer 30-minute break**.

3. Simulate Exam Conditions

- Take **full-length tests in one sitting**.

- Use a **scientific calculator** (allowed for Session 2).

- Practice **grid-in and multiple-choice questions**.

4. Review Mistakes Thoroughly

- **Keep a notebook** for incorrect answers.

- Write **step-by-step corrections** and revisit similar problems.

- Identify **patterns in mistakes** (calculation errors, concept misunderstandings).

5. Manage Test Anxiety

- Get **plenty of rest before the exam**.

- Practice **deep breathing techniques**.

- Arrive **early on test day** and stay **positive**.

Final Exam Week Plan – Last-Minute Review

Day	Task
Day 1	Review **key formulas and concepts** (Algebra Foundations & Linear Equations).
Day 2	Solve **quadratic equations, graphing & exponential problems.**
Day 3	Focus on **factoring, rational expressions, radical equations.**
Day 4	Revise **data analysis, probability, and statistics.**
Day 5	Take **a full-length timed practice test.**
Day 6	Analyze mistakes and reinforce weak topics.
Day 7	**Rest, relax, and stay confident!**

FREQUENTLY ASKED QUESTIONS

This section answers **common questions** about the **Florida Algebra 1 End-of-Course (EOC) Exam**, including exam format, scoring, preparation strategies, and more.

General Questions About the Exam

1. What is the Florida Algebra 1 End-of-Course (EOC) Exam?

The **Florida Algebra 1 EOC Exam** is a **state-administered assessment** that evaluates students' understanding of key algebraic concepts. It is required for high school graduation in Florida.

2. Who needs to take the Algebra 1 EOC Exam?

All **Florida public school students** enrolled in Algebra 1 must take the EOC exam. Private school and homeschool students may also take it if required.

3. How long is the Algebra 1 EOC Exam?

The exam is **160 minutes (2 hours and 40 minutes)**, divided into **two 80-minute sessions**. Students may use additional time if they have approved accommodations.

4. How many questions are on the test?

The test has approximately **60—66 questions**, including **multiple-choice, multi-select, grid-in, and fill-in-the-blank** questions.

5. When is the Algebra 1 EOC Exam offered?

The test is given during **three testing windows each year**:

- **Spring (April-May)**

- **Summer (June-July)**

- **Fall (September-December)**

Scoring & Passing Requirements

6. What is a passing score on the Algebra 1 EOC Exam?

A passing score is **Level 3 (497 or above)** on a **425—575 scale**.

Level	Score Range	Performance Level
Level 1	425–486	Inadequate
Level 2	487–496	Below Satisfactory
Level 3	497–517	Satisfactory (Passing Score)
Level 4	518–531	Proficient
Level 5	532–575	Mastery

7. What happens if I don't pass the exam?

Students who do not pass must:

- **Retake the exam** in a future testing window, OR

- Earn a **concordant score** on an alternate assessment such as:

  - **PSAT/NMSQT**: 430+ on Math section

  - **SAT**: 420+ on Math

  - **ACT**: 16+ on Math

8. How is the test scored?

The **Florida Department of Education (FLDOE)** scores the exam on a **computer-based system**. Scores are based on **correct answers** (no penalty for guessing).

9. How long does it take to get my results?

Scores are usually available **3–4 weeks** after the testing window closes. Your school will notify you when results are released.

Test-Taking Policies

10. Can I use a calculator on the exam?

- **Session 1: NO calculators allowed.**

- **Session 2: A scientific calculator is permitted** (provided on-screen for online testers). Graphing calculators are **not allowed**.

11. What should I bring on test day?

- School **photo ID**

- **Pencils and erasers**

- **Scratch paper** (provided by the testing center)

- **Headphones (for audio accommodations, if approved)**

12. Can I retake the test if I want a higher score?

Yes, students can **retake the Algebra 1 EOC** during the next available testing window to improve their score.

Study & Preparation Tips

13. How should I prepare for the Algebra 1 EOC Exam?

- **Follow a study plan** (1-month, 2-month, or 3-month schedule).

- **Review all key topics**: Algebra Foundations, Linear Equations, Quadratics, Polynomials, etc.

- **Take practice tests** to simulate the real exam.

- **Analyze mistakes** and focus on weak areas.

- **Use this study guide** for structured learning.

14. What are some effective test-taking strategies?

- **Manage your time:** Don't spend too long on one question.

- **Use the process of elimination:** Cross out incorrect answers.

- **Double-check grid-in responses:** Ensure accuracy in calculations.

- **Mark difficult questions:** Return to them later if time allows.

15. Where can I find additional practice materials?

- **This study guide** (includes detailed practice tests).

- **Florida Department of Education (FLDOE) website** (official sample questions).

- **Online resources**: Khan Academy, IXL, and Algebra Nation.

Exam Day Essentials

16. What should I do the night before the test?

✔ **Get a full night's sleep** (7-9 hours).
✔ **Review key formulas** and concepts briefly.
✔ **Prepare your materials** (ID, pencils, calculator).
✔ **Avoid last-minute cramming**—stay calm and confident!

17. What should I do if I feel nervous during the test?

* Take **deep breaths** to stay relaxed.

* **Skip difficult questions** and return to them later.

* Use **scratch paper** to organize your thoughts.

* **Trust your preparation**—you've got this!

Special Accommodations

18. Can students with disabilities get testing accommodations?

Yes! Students with **IEPs (Individualized Education Programs)** or **504 plans** may receive accommodations such as:

* **Extended time**

* **Small-group testing**

* **Paper-based version of the exam**

- **Use of assistive technology**

Students should **coordinate with their school counselor** well before test day.

Retesting & Future Planning

19. If I pass the Algebra 1 EOC, what's next?

After passing the Algebra 1 EOC, students typically take:

- **Geometry**

- **Algebra 2 (optional for some students)**

- **Pre-Calculus, Calculus, or other advanced math courses**

20. What should I do after taking the exam?

- Check **your score report** once results are available.

- If you passed, **celebrate your success!**

- If you didn't pass, **create a study plan** for a retake.

- **Use your Algebra 1 skills** for future math courses and standardized tests like the SAT/ACT.

1

ALGEBRA FOUNDATIONS

Real Number Properties Unveiled

Real numbers encompass all numbers on the number line, including whole numbers (0, 1, 2, 3,...), integers (..., -2, -1, 0, 1, 2,...), rational numbers (numbers that can be written as a fraction p/q where p and q are integers, and q is not zero) such as 1/2, 3/4, and -5/7, and irrational numbers (numbers that cannot be written as a fraction) such as $\sqrt{2}$, π, and e. These numbers are the foundation for all algebraic operations, and their properties allow for the manipulation of equations and expressions. Understanding the characteristics and rules governing real numbers is basic for algebra.

One of the main characteristics of real numbers is the commutative property, which states that the order in which numbers are added or multiplied does not change the result. For addition, this means that a + b = b + a. For example, if we have the expression 3 + 5, the result is 8. Similarly, if we reverse the order, 5 + 3 also equals 8. This holds true for any two real numbers being added. In multiplication, the commutative property is expressed as a × b = b × a. Take the example 2 × 4; the result is 8. If the order is reversed as 4 × 2, the result remains 8. These properties are important for simplifying and manipulating algebraic expressions, as it allows rearranging terms without altering the value of the expression.

The associative property is another key characteristic of real numbers. This property concerns how numbers are grouped in addition or multiplication. It states that the grouping of numbers does not alter the result when adding or multiplying. For addition, this is shown as (a + b) + c = a + (b + c). Consider the expression (2 + 3) + 4. First add 2 and 3 which is 5, and then add 4 which totals to 9. If the grouping is changed to 2 + (3 + 4), add 3 and 4 which is 7 and then add 2, the total is still 9. In multiplication, the property is shown as (a × b) × c = a × (b × c). Consider (2 × 3) × 4, in which first multiply 2 and 3 to equal 6, then multiply by 4 which is 24. Change the grouping to 2 × (3 × 4). Multiply 3 and 4 to equal 12, then multiply by 2 to have the same total, 24. These rules permit us to group and simplify expressions, in order to solve problems efficiently.

The distributive property explains how multiplication interacts with addition. It states that multiplying a number by a sum is the same as multiplying the number by each term in the sum and then adding the products. This is written as a × (b + c) = (a × b) + (a × c). Imagine a rectangle with a width of 'a' and a length that is the sum of 'b' and 'c' (b+c). The total area of the rectangle would be a(b+c). This is equal to the area of two smaller rectangles with a width 'a', one with a length 'b' (ab) and the other with length 'c' (ac). The total area of the two rectangles is ab + a*c. As an example, take 3 × (4 + 2). Using the distributive property, this becomes (3 × 4) + (3 × 2), which equals 12 + 6, resulting in 18. If we compute it directly as 3 × 6, the answer is also 18. The distributive property is a key tool for simplifying and solving algebraic equations, and is often used to remove parentheses from expressions.

There are special numbers known as identity elements for addition and multiplication. The additive identity is 0, meaning that any number added to 0 will remain unchanged. This is expressed as a + 0 = a and 0 + a = a. For example, 5 + 0 = 5, and -10 + 0 = -10. Zero maintains the identity of any number when used in addition. For multiplication, the identity element is 1. Any number multiplied by 1 stays the same, described as a × 1 = a and 1 × a = a. For instance, 7 × 1 = 7 and $(-2/3) \times 1 = -2/3$. These elements are the basis of many algebraic operations because they act as neutral elements in their respective operations.

Also, each real number has inverse elements for both addition and multiplication. The additive inverse of a number is the number that when added to the original number yields 0. This is also known as the opposite. The additive inverse of a number 'a' is -a, because $a + (-a) = 0$. As an example, the additive inverse of 5 is -5, since $5 + (-5) = 0$. The additive inverse of -8 is 8, since $-8 + 8 = 0$. The multiplicative inverse of a non-zero number is the number that, when multiplied by the original number, yields 1. The multiplicative inverse of 'a' is 1/a (also known as its reciprocal), where $a \times (1/a) = 1$. As an example, the multiplicative inverse of 4 is 1/4, since $4 \times (1/4) = 1$. The inverse of -2/3 is -3/2 since $(-2/3) \times (-3/2) = 1$. Zero does not have a multiplicative inverse because there is no number that when multiplied by 0 gives 1. The concept of inverses is important for solving equations, as it allows for the cancellation of terms, and isolating variables.

To check your knowledge, try the following practice problems.

Practice Problems:

   1. Identify the property that is being used in the following equation:

$(5 + 2) + 8 = 5 + (2 + 8)$

   1. Apply the distributive property to simplify the following expression:

$4(x + 3)$

   1. What is the additive inverse of -12?

   2. What is the multiplicative inverse of 5/7?

   3. What property justifies the fact that $10 + 6 = 6 + 10$?

4. Fill in the blank using the associative property:

$(3 \times 5) \times 2 = 3 \times ($ _____ $)$

1. Use the distributive property to simplify:

$9(2 + y)$

1. What is the additive identity?

2. What is the multiplicative identity?

3. Find the multiplicative inverse of $-1/3$

These rules and properties for real numbers are not just a list of rules to memorize, but instead, are the basic tools used to manipulate expressions, and to solve equations in algebra. Understanding these concepts will give the base for working with complex algebraic problems. The commutative, associative, and distributive properties, along with identity and inverse elements form a strong foundation for algebraic manipulations. In the chapters that follow, these properties will be used continuously in various mathematical contexts, and will be important for understanding and applying more advanced algebraic concepts.

Mastering PEMDAS

The order of operations, often remembered by the acronym PEMDAS, is a basic set of rules that directs how to correctly solve mathematical expressions. This roadmap ensures that calculations are done consistently and that everyone arrives at the same answer. Without a set order, the same expression could lead to different results, causing confusion and mistakes in math. PEMDAS provides this needed structure. It is an acronym that represents the sequence of operations: Parentheses, Exponents, Multiplication and Division, and Addition and Subtraction. These rules are essential in algebra and also in everyday problem-solving that involves any math.

The first step of PEMDAS is to handle anything within **Parentheses**. Parentheses or brackets, ( ), act as grouping symbols that tell us which part of the expression to solve first. When faced with an expression that contains parentheses, we do everything inside the parentheses first, as if it were a calculation on its own. This might include multiple operations inside the parentheses. When nested parentheses exist, like (( )), we work from the innermost parentheses outward, step-by-step. For example, if we see 2 × (3 + 4), we calculate what is in the parentheses first (3 + 4 = 7) and then multiply by two: 2 × 7 = 14. The parentheses create a priority, in which the math inside the parentheses is always done before operations outside of it. This use of parentheses is very important, and can drastically change an answer if not followed. Take the expression 2 × 3 + 4, if we do not use the rule, we would add first and have 2 × 7 = 14, but using PEMDAS we would multiply first and have 6 + 4 = 10.

The second step of PEMDAS involves **Exponents**. Exponents tell you how many times to multiply a number by itself. They are written as a base number with a small number, the exponent, written above and to the right. For example, in 5^2, 5 is the base and 2 is the exponent. This means 5 is multiplied by itself twice 5 × 5, and equals 25. Exponents are calculated after any operations within parentheses, but before multiplication, division, addition, or subtraction. For instance, in the expression 3 × 2^3 + 5, after any parentheses, we calculate 2^3 = 2 × 2 × 2 = 8. The expression then becomes 3 × 8 + 5. Understanding exponents is important because they are used a lot in more complex math and science problems.

After dealing with parentheses and exponents, we move to **Multiplication and Division**. These operations have the same level of priority, and are done from left to right in the expression. This means we do them in the order in which they appear. Consider the expression 10 ÷ 2 × 5. Following the left-to-right rule, we first divide 10 by 2, which equals 5, and then multiply that result by 5, which equals 25. This left-to-right rule is important, because if we multiplied first 2 × 5 = 10 and then divided, we would get 10 ÷ 10 = 1, which is not correct. Likewise, in the expression 6 × 4 ÷ 3, we would multiply first 6 × 4 = 24, and then divide by 3 which is 8. Note that division does not always take priority, even though it is listed before multiplication in the mnemonic PEMDAS; instead, we do whichever one comes first in the expression when reading left to right.

The final step of the order of operations is **Addition and Subtraction**. These, like multiplication and division, have the same priority, and are done from left to right. In an expression like 15 - 8 + 3, we first subtract 8 from 15, resulting in 7, and then add 3, which equals 10. If we did it in the opposite order, we would add 8 and 3, which is 11, and then subtract 11 from 15, which is 4, which is not correct. When we have a long expression that involves multiple additions and subtractions, we work our way through from left to right. Similar to the order between division and multiplication, addition does not always come before subtraction even though A comes before S in PEMDAS; we perform whichever of these two operations comes first when moving from left to right in the expression.

PEMDAS can be remembered through various mnemonics. A common one is "Please Excuse My Dear Aunt Sally." The first letter of each word corresponds to the operations in the correct order: Parentheses, Exponents, Multiplication and Division, Addition and Subtraction. Another way is to think of it as "Pretty Elephants March During Autumn Season". These mnemonics are helpful reminders of the order in which to tackle a mathematical expression. Visual aids such as diagrams that display each level of operations can also help. When trying to remember the rules of PEMDAS, these tools are useful to help perform the correct order of operations.

The rules of PEMDAS are used in a large range of simple to complex mathematical problems. Consider a relatively simple expression like 10 + 2 × 3 - 4. Applying

PEMDAS, we would first multiply $2 \times 3 = 6$ and have $10 + 6 - 4$. We would next add from left to right $10 + 6 = 16$, and then subtract $16 - 4 = 12$. Now consider a more complex expression, such as $(3 + 2)^2 - 4 \times 2 + 10 \div 5$. Working through the steps of PEMDAS, we first focus on the parentheses: $3 + 2 = 5$. The expression is now $5^2 - 4 \times 2 + 10 \div 5$. Next, we handle the exponent: $5^2 = 25$. Then, we perform the multiplication and division from left to right: $4 \times 2 = 8$ and $10 \div 5 = 2$. The expression is now $25 - 8 + 2$. Lastly, we do the addition and subtraction from left to right: $25 - 8 = 17$, and $17 + 2 = 19$. So, the answer to this multi-step expression is 19. These practice problems show the step-by-step process involved in using PEMDAS.

When using PEMDAS, there are common errors that students can make. One common mistake is not using the left-to-right rule for multiplication and division, and then addition and subtraction. It is important to remember that these operations have the same level of priority and are solved in the order that they appear in the equation as you read it left to right. Another frequent mistake is missing parentheses or not following the order of operations within a parenthesis. Remembering to deal with nested parentheses by starting from the inside and working outwards is very important. Also, people may make errors with exponents, not properly understanding what the exponent tells us to do, or not performing it before other operations. For example, some might try to multiply a base by its exponent, such as $2^3 = 2 \cdot 3 = 6$, when it should be $2 \cdot 2 * 2 = 8$. Awareness of these common mistakes is the first step to avoiding them. It's helpful to check each step during calculations, to make sure that no mistakes were made in any level of the process.

PEMDAS is not just an abstract mathematical rule but is instead a tool that has important applications in real-world situations. In areas of science, such as physics, the order of operations is used when working with formulas. For example, in calculating the kinetic energy of an object, which has the formula $KE = 1/2mv^2$, the exponent is calculated before multiplication. In computer programming, mathematical expressions are used a lot. It is important for these expressions to be solved correctly, so that a program will function the right way. In finance, understanding PEMDAS is needed when calculating interest or different kinds of financial formulas. Even in everyday situations like cooking, where recipes may include fractions and multiplication, PEMDAS can help in adjusting measurements correctly. Using the order of operations consistently and correctly is a very important skill, needed in various contexts.

Algebraic Expressions Decoded

Algebraic expressions are the building blocks of algebra, combining numbers and letters to represent mathematical relationships. A clear understanding of these expressions is vital for solving equations and for many different real-world applications. In these expressions, a variable is a letter or a symbol, like x or y, that represents a quantity that can change or is unknown. A constant, on the other hand, is a number that has a fixed value, such as 2, -5, or 3.14. These values do not change within a problem. Individual parts of an algebraic expression, whether they are numbers, variables, or products of numbers and variables, are called terms. For example, in the expression 3x + 5, the 3x and 5 are both terms. The term 3x includes the coefficient 3 and the variable x. An algebraic expression can contain one term, like 7, or multiple terms, like 2x + 3y - 4.

The ability to translate word problems into algebraic expressions is a vital skill. Word problems often describe relationships that must be converted into mathematical language. For instance, the phrase "five more than a number" can be turned into the expression x + 5, where x represents the unknown number. Similarly, "twice a number decreased by three" can be written as 2y - 3, where y is the unknown number. The words "product" and "quotient" mean to multiply and divide, respectively. To effectively construct algebraic expressions, it is important to carefully look at the words that describe the mathematical relationships within the problem. For example, the phrase "The sum of a number and 10 multiplied by 4" is written as 4(x+10), where the sum of x and 10 is put in parentheses, so that this sum is multiplied by 4. Without the parentheses, 4x+10 would mean that only x is multiplied by 4, which is not what the problem describes.

After creating an algebraic expression, the next step is substituting values for the variables, so you can evaluate the expression to get a numerical answer. This means replacing the variables with specific numbers and following the order of operations, PEMDAS, to get the final answer. For example, if we have an expression 2a + 3b, and a = 4 and b = 2, we substitute these values into the expression 2(4) + 3(2). We then do the multiplication first: 2(4) = 8 and 3(2) = 6. Lastly, we add, 8 + 6 = 14. So, when a equals 4 and b equals 2, the value of the expression 2a + 3b is 14. The ability to substitute values into expressions and evaluate them is important in algebra, and also for practical problems involving formulas. Consider the formula for area of a triangle: A = (1/2)bh, where b stands for the base of the triangle and h is the height. If we know that the base is 6 and the height is 4, we substitute these values to find the area: A = (1/2)(6)(4). We perform the multiplication from left to right to get A = (1/2)(24) and then do the last step A = 12.

When solving problems involving algebraic expressions, a systematic approach is very helpful. First, identify what the problem gives you and what you need to find out. Make a list of all known values and carefully read the word problem to find out what the question is asking for. Second, decide which letters will represent the unknown values. The letters you choose will be your variables. Third, translate the problem into an algebraic expression that shows the relationship between the variables and constants. Fourth, evaluate the expression by substituting in any known values and simplifying it. This approach helps reduce errors and keeps the process organized, which makes complex problems easier to handle. For example, suppose we have a problem that says, "A taxi charges a \\$3 initial fee, and an additional \\$2 for every mile traveled. If a person travels 10 miles, how much will the taxi charge?" First, we identify what is known: the initial fee is \\$3, the cost per mile is \\$2, and the distance is 10 miles. Let m represent the number of miles traveled. The expression is 3 + 2m. Now, substitute m = 10: 3 + 2(10). Doing the multiplication first, we have 3 + 20, and then the addition 23. So, the total taxi charge will be \\$23.

Working with word problems can sometimes be difficult, and requires both creativity and precision when translating from the word problem into algebraic language. Consider a problem such as this: "A store sells shirts for \\$15 each and pants for \\$25 each. If someone buys three shirts and two pants, what is the total cost before tax?" The first step is to define variables for the number of shirts and pants purchased, like s for the number of shirts and p for pants. The cost for the shirts will be 15s, and the cost for the pants is 25p. To find the total cost before tax, we use the expression 15s + 25p. Then, substitute the numbers given in the problem: 15(3) + 25(2). Evaluate the multiplication first, 45 + 50, and then the addition, 95. So, the total cost is \\$95. These word problems show that by first translating the problem carefully, an algebraic expression can be made to represent the situation, and then we can substitute and evaluate to get the answer.

In some cases, you may have to construct an expression involving multiple variables and steps. Consider this problem: "A rectangular garden has a length that is 5 feet longer than its width. If the width is w feet, what is an expression for the perimeter of the garden?" The width is w, and the length is w + 5. Since a rectangle has two pairs of equal sides, the perimeter is twice the sum of the length and the width, or 2(w + (w + 5)). We simplify the inside: 2(2w + 5). Then we distribute the 2: 4w + 10. So, the perimeter is represented by 4w + 10 feet. If the width is 10 feet, then we can substitute and get 4(10) + 10 = 50 feet. Algebraic expressions can also be used to model relationships in formulas. Consider this word problem: "A train travels at a speed of s miles per hour. How far will it travel in 3 hours and 30 minutes? Express the distance it travels as a function of s." Distance is speed multiplied by time. The time is 3 hours and 30 minutes, which can be expressed as

3.5 hours. So, the distance is 3.5s miles. If the train's speed is 60 miles per hour, the train will travel 3.5 * 60 = 210 miles. These examples of more complex problems show the importance of developing skills in creating algebraic expressions and working with variables.

Checking your work is an important part of solving algebraic expression problems. One way to check your work is to substitute the values again in a different order, to verify the answer. If possible, look at the problem logically, to see if the answer you got makes sense in the context of the word problem. For example, if you have a problem about measuring the perimeter of a field, you would expect your answer to be in some length unit, such as feet or meters. Another way to check is to work through the problem backwards, if possible. If you are starting with an expression and substituting a number, try working backwards with the answer to see if you can get back the starting expression. In more difficult problems, you may need to ask someone else, such as a tutor or a teacher, to check your answer and explain to you any errors.

The skills used to work with algebraic expressions have many practical uses. In science, many formulas used to describe physical phenomena involve variables and algebraic expressions. In physics, for example, the relationship between distance, speed, and time is shown with the formula $d = rt$, where d is distance, r is the rate of speed, and t is time. In engineering, algebraic expressions are used to design and build structures and machines by modeling different forces and physical laws. In everyday life, the use of algebraic expressions can be seen in budgeting, shopping, cooking, and in figuring out any kind of measurement. For example, you can use algebra to calculate the cost of buying multiple items in a store, as shown in the word problem with shirts and pants earlier. These mathematical skills are used a lot, both in academia and in real-world situations, which makes it important to develop these skills. The understanding of algebraic expressions makes solving complex problems much easier.

Expression Simplification Techniques

Simplifying algebraic expressions involves reducing them to their most basic form, which makes them easier to understand and work with. This process usually involves combining like terms and applying the distributive property. The first step is to know what like terms are. Like terms are terms that contain the same variables raised to the same powers. For instance, 3x and 5x are like terms because they both have the variable x raised to the power of one. On the other hand, 3x and $5x^2$ are not like terms because the variable x is raised to different powers, one and two, respectively. Similarly, 3x and 5y are not like terms because they have different variables. Constant terms, such as 4 and -7, are also considered like terms.

To combine like terms, add or subtract their numerical coefficients while keeping the variable part the same. For instance, to simplify 3x + 5x, we add the coefficients 3 + 5 = 8, which gives 8x. Similarly, 7y - 2y becomes (7-2)y, which simplifies to 5y. The coefficients are the numerical part of the terms; in 3x, the coefficient is 3. If a term appears by itself without a written coefficient, such as x, the coefficient is understood to be 1. Thus, x is the same as 1x. For example, if you have an expression such as x + 4x - 2x, we combine the coefficients 1 + 4 - 2 = 3, giving us 3x. Terms are grouped together and combined to reduce the number of terms, thus simplifying the expression. If you have a mixture of different variable terms, you combine each group separately. For example, given 2a + 3b - a + 5b, we combine the a terms to get 2a - a = 1a or a, and we combine the b terms to get 3b + 5b = 8b. The resulting simplified expression is a + 8b. Remember, terms with different variables or different powers can not be combined using addition or subtraction.

The distributive property is a key technique used when simplifying algebraic expressions involving parentheses. It states that multiplying a sum (or difference) by a number is the same as multiplying each term in the sum (or difference) individually by that number. In general, a(b + c) = ab + ac. For instance, in the expression 3(x + 2), the 3 is distributed to both x and 2, resulting in 3x + 6. Another example could be 5(2y - 4), where 5 is multiplied by 2y and 4, giving 10y - 20. If there is a negative sign in front of the parentheses, it can be thought of as multiplying by -1. For example, in -(3z + 5), the entire expression inside the parentheses is multiplied by -1, which makes -3z - 5. Similarly, -(4x - 6) becomes -4x + 6. Remember that multiplying a negative by a negative gives a positive. When distributing, pay careful attention to signs.

A step-by-step approach is very helpful for simplifying algebraic expressions: First, identify all the individual terms in the expression. Look for terms separated by plus or minus signs, both inside and outside parentheses. Second, group like terms together, which may require using the commutative property of addition, which says you can add numbers in any order. So, 2x + 3y + 4x - y can be rewritten as 2x + 4x + 3y - y to group the x's and the y's. Third, combine the like terms by adding or subtracting the numerical coefficients, as explained earlier. Finally, make sure the expression has been simplified as much as possible, that there are no more like terms to combine, and all parentheses have been cleared by distributing.

Let's practice a few examples: To simplify 5x + 3 + 2x - 1, we identify the terms: 5x, 3, 2x, and -1. We group the like terms together: 5x + 2x + 3 - 1. We then combine the like terms: (5+2)x + (3-1) = 7x + 2. Therefore the simplified expression is 7x + 2. For an example with the distributive property, let's simplify 2(3y - 4) + 5y. First, we distribute the 2: 2 3y - 2 4 + 5y = 6y - 8 + 5y. Then we group like terms: 6y + 5y - 8. We combine the y terms: 11y - 8. The simplified expression is 11y - 8. For a more complicated example with negative signs, consider -3(2a - 5) + 4a - 10. First, distribute the -3: -32a -3-5 + 4a - 10 = -6a + 15 + 4a - 10. Group like terms: -6a + 4a + 15 - 10. Combine like terms: (-6 + 4)a + (15-10) = -2a + 5. The simplified expression is -2a + 5.

When working with more complicated expressions with multiple variables, the same steps are followed. Consider simplifying 3x + 2y - 4x + 5z - 2y + z. First, group like terms together: 3x - 4x + 2y - 2y + 5z + z. Then, combine like terms: (3-4)x + (2-2)y + (5+1)z. The simplified result is -x + 0y + 6z, which we can rewrite as -x + 6z because 0y is just zero. It is important to be careful when combining negative terms and terms with negative coefficients, such as in the example 7a - 8b + 4a - 2b, which groups as 7a + 4a - 8b - 2b. Combining like terms gives us 11a - 10b. Pay careful attention to the signs of the coefficients, using the rules of positive and negative numbers.

One common mistake is to combine terms that are not like terms. For example, in $2x + 3x^2$, it is incorrect to combine the 2x and $3x^2$. Because the variable terms have different powers, these terms cannot be combined. Another mistake is forgetting to distribute a number across all terms in parentheses, or making errors with the signs when distributing. For example, in the expression -(a - b), the entire expression is multiplied by -1, making it -a + b. It is important to remember to apply the negative sign to each term in the parentheses. Another mistake is not combining like terms correctly. For instance, in the expression 5x + 2 + 3x - 1, combining the x terms separately from the constant terms is important. You add 5x + 3x = 8x, and 2 - 1 = 1, making the simplified expression 8x + 1, not 10x. To avoid this, always carefully look at the terms, and make sure to combine only the ones that are truly like terms. To prevent these kinds of errors, use the steps of identifying each term, grouping like terms, combining the coefficients, and simplifying in order. Double checking the distribution and all signs is also a useful method for preventing errors.

When you have a complicated problem, sometimes it helps to rewrite it to make it easier to handle, such as grouping all terms with the same variable together, or writing out all of the multiplication to make it clear what to do, step by step. For complex problems, it may be helpful to rewrite the problem and distribute or rearrange terms, keeping all the terms clearly written, without skipping any steps.

Then, simplify slowly, one part at a time. It can be very helpful to visually track terms using different colors or shapes, underlining all the x's in one color, the y's in a different color, and so on. This helps make sure that you don't miss any terms or mistakenly combine unlike terms. Use a systematic approach, which will help you not lose track of what you are doing. Write each term in its simplest form before combining them. The goal is to make complex problems manageable by breaking them into simpler steps, which reduces the chance of errors.

Mastery of simplification helps with more advanced math problems. It is necessary for solving equations, finding the solution to complicated systems of equations, and also for dealing with functions. The ability to quickly simplify an algebraic expression to its simplest form will be very helpful for working more complex problems.

Solving Equations and Inequalities

Solving equations and inequalities is a fundamental skill in algebra, allowing us to find unknown values and understand relationships between quantities. The basic goal in solving an equation is to isolate the variable, meaning to get the variable by itself on one side of the equal sign. This is done by using inverse operations, which "undo" the operations in the equation. For example, if a number is being added to the variable, we subtract that number from both sides of the equation. If a variable is being multiplied by a number, we divide both sides by that number. The key principle in solving equations is maintaining balance. Whatever operation is done on one side of the equation, must also be done on the other side to keep the equation true. The equal sign represents a balance, and all steps taken must keep that balance intact.

When dealing with one-step equations, isolating the variable requires a single operation. Consider the equation $x + 5 = 12$. To solve for x, we need to undo the addition of 5. The inverse operation of addition is subtraction, so we subtract 5 from both sides of the equation. This gives us x + 5 - 5 = 12 - 5, which simplifies to x = 7. To check if our answer is correct, we substitute x = 7 back into the original equation: 7 + 5 = 12. Since 12 = 12, our answer is correct. Another example is the equation y - 3 = 8. To isolate y, we add 3 to both sides: y - 3 + 3 = 8 + 3, which gives y = 11. Checking the answer, we substitute y = 11 into the original equation: 11 - 3 = 8, and 8=8, so the solution is correct. For equations involving multiplication, for example 3a = 15, we divide both sides by 3: 3a / 3 = 15 / 3, which simplifies to a = 5. The check is 3 5 = 15, and 15=15, showing the result is correct. For an equation such as b / 4 = 2, we multiply both sides by 4: (b / 4) 4 = 2 * 4, which means b = 8. Checking, we do 8/4 = 2, and 2=2, which is correct. These examples show how inverse operations are used to isolate the variable, solving one-step equations.

Two-step equations require two operations to isolate the variable. For instance, take the equation 2x + 3 = 11. Here, x is being multiplied by 2 and then 3 is added. To solve, we must undo the addition first, using the inverse operation of subtracting 3 from both sides of the equation, which gives 2x + 3 - 3 = 11 - 3. This simplifies to 2x = 8. Now, x is being multiplied by 2, so we divide both sides by 2: 2x / 2 = 8 / 2, resulting in x = 4. To verify the solution, substitute x=4 back into the original equation: 2 4 + 3 = 11. Evaluating, we get 8 + 3 = 11, which simplifies to 11 = 11, showing the answer is correct. Consider another example, (y / 4) - 2 = 3. The first step is to undo the subtraction of 2 by adding 2 to both sides: (y / 4) - 2 + 2 = 3 + 2, which simplifies to y / 4 = 5. The next step is to multiply both sides by 4: (y / 4) 4 = 5 * 4, which leads to y = 20. Checking, (20 / 4) - 2 = 3. Evaluating, 5 - 2 = 3, which gives 3 = 3, verifying that the answer is correct. When solving two-step equations, it is important to follow the correct order of inverse operations, undoing additions and subtractions first, followed by multiplications and divisions.

Some equations may have variables on both sides of the equal sign. For example, consider 5a + 4 = 2a + 10. To solve, we want to get all the terms containing a on one side and the constant terms on the other. We could first subtract 2a from both sides, this gives us 5a + 4 - 2a = 2a + 10 - 2a, which simplifies to 3a + 4 = 10. Then, we subtract 4 from both sides, 3a + 4 - 4 = 10 - 4, which simplifies to 3a = 6. Finally, we divide both sides by 3: 3a / 3 = 6 / 3, which gives a = 2. To verify, we substitute a=2 into the original equation: 5 2 + 4 = 2 2 + 10. Evaluating, we get 10 + 4 = 4 + 10, which simplifies to 14 = 14, confirming the answer. Another example is the equation 7x - 3 = 4x + 9. We start by subtracting 4x from both sides: 7x - 3 - 4x = 4x + 9 - 4x, simplifying to 3x - 3 = 9. Then, we add 3 to both sides: 3x - 3 + 3 = 9 + 3, which simplifies to 3x = 12. Finally, we divide both sides by 3: 3x / 3 = 12 / 3, which gives x = 4. We verify by substituting x=4 back into the original equation: 7 4 - 3 = 4 4 + 9. This evaluates to 28 - 3 = 16 + 9, and then to 25 = 25. Solving equations with variables on both sides often requires a few extra steps, but the principle is the same, always using inverse operations to isolate the variable while maintaining balance on each side of the equation.

In addition to equations, we also need to solve inequalities. Inequalities use symbols such as < (less than), > (greater than), ≤ (less than or equal to), and ≥ (greater than or equal to). Solving inequalities is similar to solving equations, but there is one main difference: when multiplying or dividing both sides of the inequality by a negative number, we must flip the inequality sign. For example, if we have -x < 4, we divide both sides by -1, and since we are dividing by a negative number, we must flip the sign: x > -4. Consider the inequality x + 3 < 7. To isolate x, we subtract 3 from both sides: x + 3 - 3 < 7 - 3, resulting in x < 4. This indicates that any value of x that is less than 4 will satisfy the inequality. For a two-step inequality, such as 2y - 1 ≥ 5, we start by adding 1 to both sides: 2y - 1 + 1 ≥ 5 + 1, which gives 2y ≥ 6. Then, we divide both sides by 2: 2y / 2 ≥ 6 / 2, leading to y ≥ 3. This means that y can be any value that is greater than or equal to 3. Another example is -3a + 5 > 14. We start by subtracting 5 from both sides: -3a + 5 - 5 > 14 - 5, which simplifies to -3a > 9. Since we are going to divide by a negative number, -3, we must flip the inequality sign, giving a < -3.

The solutions to inequalities can be represented graphically on a number line. For the inequality x < 4, we draw a number line, mark the point 4, and draw an open circle at 4 to show that 4 is not included in the solution. Then we draw an arrow to the left of 4 to indicate that all values less than 4 are solutions. For the inequality y ≥ 3, we draw a number line, mark the point 3, and draw a closed circle at 3 to show that 3 is included in the solution. Then we draw an arrow to the right of 3 to indicate that all values greater than 3 are solutions. For inequalities such as a < -3, draw a number line and mark the point -3. Draw an open circle at -3, and draw an arrow to the left to show that all values less than -3 are solutions. The open circle shows that -3 is not part of the answer, and the arrow shows the solution set for the inequality.

Equations and inequalities are applied in many real-world situations. For example, consider a scenario where a student is saving money. If they start with $20 and save $5 each week, we can represent the total amount of money saved after w weeks as 20 + 5w. If the student wants to save at least $100, we can write an inequality: 20 + 5w ≥ 100. To find how many weeks are needed, we solve the inequality: 5w ≥ 80, and w ≥ 16. This means the student needs at least 16 weeks to save $100. Another scenario involves calculating the cost of renting a car. If the rental fee is $30 per day plus a one-time fee of $50, the total cost for d days can be written as 30d + 50. If a person has a budget of $200, the inequality would be 30d + 50 ≤ 200. Solving this we have: 30d ≤ 150, and d ≤ 5. This means the person can rent the car for a maximum of 5 days within their budget. These examples show how equations and inequalities are essential for problem-solving in real-world scenarios.

Solving equations and inequalities is a basic part of algebra, science, and problem-solving. The methods involve using inverse operations, maintaining balance, and handling inequalities properly by flipping the sign when multiplying or dividing by a negative. Understanding these methods is key for working on complex mathematical tasks.

Absolute Value Essentials

Absolute value represents the distance a number is from zero on the number line, always expressed as a non-negative value. The absolute value of a number, denoted by |x|, signifies the magnitude of that number without regard to its sign. For instance, the absolute value of 5, written as |5|, is 5, while the absolute value of -5, written as |-5|, is also 5. This definition means that absolute value functions as a method for finding the numerical distance of a number from zero, ignoring whether the number is to the left or right of zero on the number line.

Calculating the absolute value of a number is straightforward. If the number is positive or zero, its absolute value is the number itself. For example, |10| = 10, and |0| = 0. If the number is negative, its absolute value is the opposite of that number, effectively making it positive. For instance, |-7| = 7, and |-2.5| = 2.5. This always-positive nature is a fundamental characteristic of absolute value, which is why the result is never a negative number. This is because distance is a concept that is always measured as a non-negative value. Therefore, whether a number is situated to the right or left of zero, its distance from zero is always positive.

When solving equations involving absolute values, it's essential to consider that the expression inside the absolute value bars could be either positive or negative, and still yield the same absolute value. For example, if we have |x| = 3, there are two possibilities: x could be 3 or -3, since both |3| and |-3| equal 3. To solve absolute value equations, we consider both cases to ensure that we identify all possible solutions. Consider the equation |x - 2| = 4. This equation requires us to determine all values of x for which the distance between x and 2 is equal to 4. We solve two separate equations: x - 2 = 4 and x - 2 = -4. Solving the first equation, x - 2 = 4, we add 2 to both sides, giving x = 6. Solving the second equation, x - 2 = -4, we add 2 to both sides, giving x = -2. Therefore, the solutions to |x - 2| = 4 are x = 6 and x = -2. It is essential to substitute each solution back into the original equation to confirm its accuracy.

A slightly more involved problem includes an absolute value expression on one side of an equation such as $|2x + 1| = 7$. To solve, we set up two equations. The first equation is $2x + 1 = 7$, and the second is $2x + 1 = -7$. Solving the first equation, $2x + 1 = 7$, we subtract 1 from both sides, getting $2x = 6$, and then divide by 2 which leads to x = 3. Solving the second equation, $2x + 1 = -7$, we subtract 1 from both sides, resulting in $2x = -8$, and then divide by 2 which leads to x = -4. Substituting x=3 into the original equation, we have $|2(3) + 1| = |6 + 1| = |7| = 7$, which confirms the solution. Substituting x=-4 into the original equation, we have $|2(-4) + 1| = |-8 + 1| = |-7| = 7$, which also confirms the solution. Thus, the solutions to $|2x + 1| = 7$ are x = 3 and x = -4.

Graphing absolute value functions reveals their unique V-shape. The basic absolute value function, $y = |x|$, forms a V shape with the vertex at the origin (0,0), because values of x, whether positive or negative result in positive values of y. When the absolute value function is altered, the graph transforms, such as with $y = |x - h| + k$, where h shifts the vertex horizontally and k shifts it vertically. For example, $y = |x - 2| + 3$ is shifted 2 units to the right and 3 units upward, compared to $y = |x|$. The vertex of $y = |x - 2| + 3$ is at (2,3). Similarly, the function $y = |x+1| - 4$ has its vertex at (-1,-4). The slope of the two sides of the V shape in $y = |x|$ is 1 for x greater than 0, and -1 for x less than 0. When there is a coefficient, a, in front of the absolute value such as $y = a|x|$, this influences how steep the sides of the graph will be. If a is a positive number greater than 1, the sides will be steeper than $y=|x|$ and if the positive number is less than 1, the sides will be shallower than $y=|x|$. If a is negative, this creates a downward opening V shape. Graphing helps to visualize how different changes to the absolute value function transform its shape and location.

Solving more complex absolute value equations involves multiple steps, such as $|3x - 5| + 2 = 9$. First, we isolate the absolute value term by subtracting 2 from both sides: $|3x - 5| = 7$. Then we split this into two separate equations: $3x - 5 = 7$ and $3x - 5 = -7$. Solving the first equation, $3x - 5 = 7$, we add 5 to both sides, giving $3x = 12$, and then divide by 3, which leads to $x = 4$. Solving the second equation, $3x - 5 = -7$, we add 5 to both sides, giving $3x = -2$, and then divide by 3, which leads to $x = -2/3$. Therefore, the solutions to $|3x - 5| + 2 = 9$ are $x = 4$ and $x = -2/3$. Substituting $x=4$ back into the equation, we have $|3(4) - 5| + 2 = |12 - 5| + 2 = |7| + 2 = 7 + 2 = 9$. Substituting $x = -2/3$ back into the equation, we have $|3(-2/3) - 5| + 2 = |-2-5| + 2 = |-7| + 2 = 7 + 2 = 9$. Thus, both solutions are correct.

A challenging problem may involve nested absolute values, such as $||x - 3| - 2| = 1$. Here, we tackle the outer absolute value first, meaning we consider two cases: $|x - 3| - 2 = 1$ and $|x - 3| - 2 = -1$. In the first case, add 2 to both sides to obtain $|x-3|=3$. This gives two possibilities: $x-3=3$ which leads to $x=6$, and $x-3=-3$ which leads to $x=0$. In the second case, add 2 to both sides, to obtain $|x-3|=1$. This gives two further possibilities: $x-3=1$, which leads to $x=4$, and $x-3=-1$, which leads to $x=2$. Thus the equation $||x - 3| - 2| = 1$ has four solutions, $x=6$, $x=0$, $x=4$, and $x=2$. To verify, substitute these four solutions into the original equation. When $x=6$, $||6 - 3| - 2| = ||3| - 2| = |3 - 2| = |1| = 1$, which is correct. When $x=0$, $||0 - 3| - 2| = ||-3| - 2| = |3 - 2| = |1| = 1$, which is also correct. When $x=4$, $||4 - 3| - 2| = ||1| - 2| = |1 - 2| = |-1| = 1$, which is correct. When $x=2$, $||2 - 3| - 2| = ||-1| - 2| = |1 - 2| = |-1| = 1$, which confirms all four solutions. These examples demonstrate the process of solving more difficult absolute value problems.

Real-world applications of absolute value are abundant, primarily in contexts involving measurement, distance, and error analysis. In physics, calculating the magnitude of a vector, such as velocity or force, involves using absolute value to determine its numerical value regardless of its direction. For example, if an object's velocity is -5 meters per second, its speed, the magnitude of velocity, is $|-5| = 5$ meters per second. In engineering, absolute value can be used to specify tolerance levels in manufacturing, indicating how much variation is allowed from a target measurement without causing an error. A tolerance of plus or minus 0.01 inches from a 5 inch rod could be shown as $|5-x| \leq 0.01$, where x is the actual size. In daily life, consider measuring the distance between two points on a map; absolute value can be used to find the distance regardless of the order in which the coordinates are subtracted. If point A is at 2 miles and point B is at 7 miles, the distance from A to B, $|7-2| = 5$ miles is the same as the distance from B to A, $|2-7|=|-5|=5$ miles.

Common misconceptions about absolute value often center on its relationship to the sign of a number. Many might think that absolute value simply flips the sign of any number, when it only affects negative numbers. When a number is positive or zero, its absolute value remains the same. Another confusion is in solving absolute value equations, where learners sometimes forget to consider both positive and negative cases. When solving $|x-a| = b$, for example, it is imperative to solve for $x-a=b$ and $x-a=-b$. Failing to consider both cases can lead to incomplete answers. Understanding that absolute value represents a distance from zero, and that distance is always positive, helps to clarify these points of confusion. In scenarios where absolute values are combined with variables, such as $|x| + x$, understanding when the absolute value has no impact on positive values of x but reverses negative values is essential to accurately solve equations.

Absolute value is essential for a wide range of calculations in mathematics and the sciences. Its applications extend from the simple determination of distance to sophisticated calculations in error analysis and vector magnitude. In physics and engineering, it is the basic principle when finding magnitudes of forces, velocities and tolerances. Its ability to show the magnitude of a value without considering its direction or sign means that absolute value is vital for solving a wide array of problems.

2

LINEAR EQUATIONS AND FUNCTIONS

Function Fundamentals: Notation and Relationships

A function is a mathematical relationship where each input is associated with exactly one output. This concept is fundamental to algebra and much of higher mathematics. It describes how one quantity depends on another. Function notation provides a clear and concise way to express these relationships, using symbols to represent inputs and outputs. The most common notation is $f(x)$, where f is the name of the function, and x is the input, also known as the independent variable. The output, or dependent variable, is represented by $f(x)$. Think of a function as a machine: you put something in (the input), and it gives you something out (the output) based on a specific rule. This rule is what defines the function.

The input, typically represented by x, is the value that we provide to the function. It's also called the independent variable because its value is not determined by anything within the function itself. The output, represented by $f(x)$, is the result of applying the function's rule to the input value. It is called the dependent variable because its value depends on the chosen input. The domain of a function is the set of all possible input values (all the x values) for which the function is defined. For example, if a function involves a square root, its domain would not include negative numbers because the square root of a negative number is not a real number. The range of a function is the set of all possible output values (all the f(x) values) that the function can produce. Determining the domain and range is essential for understanding the limits and capabilities of a function.

A function is valid only if each input has exactly one output. This can be examined using the vertical line test on graphs or by checking that no input in a table or list corresponds to more than one output. When graphed on a coordinate plane, if any vertical line intersects the graph at more than one point, the relation is not a function. In table form, if an x value appears more than once, and is associated with different y values, it is not a function.

Let's consider how to determine the domain and range of a function presented in different ways. When given a set of points (such as in a table), the domain is simply the set of all the unique x-values, and the range is the set of all unique y-values. For example, given the set of points {(1,2), (2,4), (3,6)}, the domain would be {1, 2, 3}, and the range would be {2, 4, 6}. When given a graph, the domain is all the x-values covered by the graph, and the range is all the y-values covered. If the graph is a line segment, the endpoints of the segment need to be taken into consideration for determining the domain and range. If the graph is a continuous curve, it will cover an interval of x and y values that should be considered when determining domain and range. When given an equation of a function, we need to consider what inputs produce real number outputs. Sometimes, there might be restrictions, such as avoiding division by zero or taking the square root of a negative number. In the function $f(x) = 1/x$, the domain excludes 0 because division by zero is undefined. For $f(x) = \sqrt{x}$, the domain is all non-negative numbers, because the square root of a negative number is not a real number. In the case of linear functions, unless explicitly stated otherwise, the domain and range are typically all real numbers.

The input-output relationship is what the function is all about, connecting what we put into it with what we get out. It helps us see how changes in one variable affect another variable, making functions valuable in problem-solving.

Understanding this relationship lets us predict results, model real-world events, and make informed choices. For example, if you have the function $d(t) = 60t$, which represents the distance d traveled at 60 miles per hour in t hours, inputting $t = 2$ will produce an output of $d(2) = 120$. This illustrates how, given an input (time), we calculate an output (distance), clearly demonstrating the input-output relationship.

Let's explore more real-world examples. If a car is traveling at a constant speed of 50 miles per hour, the distance d covered in t hours is given by the function $d(t) = 50t$. Here, t is the input (time), and $d(t)$ is the output (distance). If $t = 3$, then $d(3) = 50 \cdot 3 = 150$, meaning the car travels 150 miles in 3 hours. Similarly, consider the cost calculation with variable pricing, where a taxi charges a base fare of \$5 plus \$2 per mile. The total cost c for m miles is given by $c(m) = 5 + 2m$. If you travel 10 miles, then $c(10) = 5 + 2 \cdot 10 = 25$, making the fare \$25. For temperature conversions, the function $F(C) = (9/5)C + 32$ converts degrees Celsius (C) to degrees Fahrenheit (F). If the temperature is 20°C, then $F(20) = (9/5) \cdot 20 + 32 = 68°F^*$. All these examples illustrate functions in action, each one having a clearly defined input-output relationship.

To further solidify understanding, here are some practice problems. First, let's practice evaluating functions. Consider the function $g(x) = 2x^2 - 3x + 1$. What is $g(4)$? To find this we substitute 4 for every x in the expression, which yields $g(4) = 2(4)^2 - 3(4) + 1 = 2(16) - 12 + 1 = 32 - 12 + 1 = 21$. Similarly, if given $h(x) = 5 - x / 2$, what is $h(10)$? Substituting yields $h(10) = 5 - 10 / 2 = 5 - 5 = 0$. These examples illustrate how to take a specific input value and get an output value according to the function's rule.

Next, let's consider how to tell if a relation represents a function. Given the points {(1,3), (2,5), (3,7), (1,4)}, is this a function? No, it is not because the input value of $x = 1$ corresponds to two different output values, $y = 3$ and $y = 4$. Now, let's consider a table:

x	y
-2	4
-1	1
0	0
1	1
2	4

Does this table represent a function? Yes, it does. Notice that every x-value has exactly one corresponding y-value. Even though y values repeat, what is important is that x values don't repeat with a different y value. How about the relation defined by $y = \pm\sqrt{x}$? Is this a function? This is not a function. For any positive x, there would be two corresponding y-values: a positive and a negative square root. For example, for x=4, we would get both y=2 and y=-2. The function $y = x^2$, however, is a valid function because each x-value has exactly one corresponding y value.

Finally, let's work on interpreting function notation in context. Suppose the function p(h) = 15h represents the pay p in dollars for working h hours. What does p(8) = 120 mean? This means that for working 8 hours, the pay is $120. What if p(h) = 200? What does it mean, and how would you solve it? p(h) = 200 means that we need to find the number of hours h where the pay is $200. Substituting in the function gives us 200 = 15h, which gives us h = 200 / 15 = 13.33 hours, roughly.

Functions are an essential tool for modeling real-world relationships because they provide a way to predict and understand the interaction between different quantities. Understanding function notation and being able to use it confidently is essential for succeeding in higher-level math. They allow us to represent relationships between variables in a clear and concise way, which is fundamental to problem-solving. Being able to translate real-world situations into mathematical functions enables effective planning and analysis.

Graphing Linear Equations Masterclass

Graphing linear equations is a visual way to understand the relationship between two variables, often represented as _x_ and _y_, where the relationship forms a straight line. To start, the foundation is the coordinate plane, also called the Cartesian plane. It is formed by two perpendicular number lines: the horizontal _x_-axis and the vertical _y_-axis, intersecting at a point called the origin, which is at the coordinate (0,0). Every point on the coordinate plane is identified by a unique pair of coordinates, written as (x, y), where the _x_ value represents the horizontal position relative to the origin, and the _y_ value represents the vertical position. For example, the point (3, 2) is located three units to the right of the origin and two units above the origin. The coordinate plane is divided into four quadrants by the two axes, which are numbered using roman numerals counter-clockwise, starting with the top-right quadrant as quadrant I. Plotting points accurately is essential for accurately graphing linear equations, and involves moving left or right according to the _x_ value and up or down according to the _y_ value from the origin.

Key points on a linear graph are the _x_-intercept and the _y_-intercept. The _x_-intercept is the point where the line crosses the _x_-axis; at this point, the _y_ value is zero. To find the _x_-intercept, you set _y_ to zero in the equation and solve for _x_. The _y_-intercept is the point where the line crosses the _y_-axis; at this point, the _x_ value is zero. To find the _y_-intercept, you set _x_ to zero in the equation and solve for _y_. These intercepts provide a clear view of where the line crosses each axis and help to locate the line accurately on the coordinate plane.

When combined with other points, the intercept points help to define a linear graph. For example, consider the equation _y = 2x + 4_. To find the _x_-intercept, set _y = 0_: _0 = 2x + 4_. Solving for _x_ gives _x = -2_, so the _x_-intercept is (-2, 0). To find the _y_-intercept, set _x = 0_: _y = 2(0) + 4_, which gives _y = 4_, so the _y_-intercept is (0, 4). These two points give us two key reference points for graphing the line.

The slope of a line, often denoted by the letter _m_, is a measure of its steepness and direction. It quantifies the rate at which the _y_-value changes with respect to the _x_-value. To calculate the slope given two points, (x_1, y_1) and (x_2, y_2), the formula is: $m = (y_2 - y_1) / (x_2 - x_1)$. The slope represents the "rise over run," where the rise is the change in _y_ and the run is the change in _x_. For example, if we have the points (1, 3) and (3, 7), the slope would be calculated as m = (7 - 3) / (3 - 1) = 4 / 2 = 2. This means that for every 1 unit increase in _x_, the value of _y_ increases by 2 units.

The sign of the slope tells us the direction of the line. A positive slope indicates that as _x_ increases, _y_ also increases, which is a line going up and to the right. A negative slope indicates that as _x_ increases, _y_ decreases, which is a line going down and to the right. A slope of zero means that the line is horizontal. This would be the case in an equation such as _y = 5_, where for any x-value, the _y_ is always 5. A vertical line, such as _x = 3_, would have an undefined slope because the change in _x_ is zero. If you were to put zero in the denominator of the slope formula, you would get a situation where the value is undefined. Understanding the sign and magnitude of the slope allows you to understand how the dependent variable changes as the independent variable changes. The slope is a constant value for any given linear equation.

To graph a linear equation effectively, one technique is to create a table of values, also known as a t-chart. This involves choosing a range of _x_ values, substituting them into the equation, and calculating the corresponding _y_ values. These points (x, y) are then plotted on the coordinate plane. For example, consider the equation _y = -x + 5_. Choose a few x-values, like -1, 0, 1, and 2. When x = -1, y = -(-1) + 5 = 6, giving the point (-1, 6). When x = 0, y = -(0) + 5 = 5, giving the point (0, 5). When x = 1, y = -(1) + 5 = 4, giving the point (1, 4). When x = 2, y = -(2) + 5 = 3, giving the point (2, 3). Once these points are calculated, plot them on the coordinate plane. Then, draw a straight line that passes through these points.

When plotting points, ensure you move along the _x_-axis first, then along the _y_-axis. Use a ruler or straight edge to draw the line accurately. A consistent scale on the axes will prevent distortion. Be aware of how your scale impacts the steepness of the line in the visual. For instance, if one axis has a scale of 1 unit per space and another axis has a scale of 5 units per space, the line might look steeper or flatter than it is. When drawing the line, make sure it extends beyond the plotted points and includes arrows on both ends to show that the line goes on infinitely.

Linear equations come in various forms, including slope-intercept form, standard form, and point-slope form. The slope-intercept form is _y = mx + b_, where _m_ is the slope and _b_ is the _y_-intercept. For example, the equation _y = 3x - 2_ is in slope-intercept form, where the slope is 3 and the _y_-intercept is -2. The standard form is _Ax + By = C_, where _A_, _B_, and _C_ are constants. For example, the equation _2x + 3y = 6_ is in standard form. To graph an equation in standard form, you can either find the intercepts, or rearrange it into slope-intercept form. The point-slope form is _y - y_1 = m(x - x_1)_ where _m_ is the slope and (x_1, y_1) is a known point on the line. For example, if a line has a slope of -2 and passes through the point (1, 4), the point-slope form of the equation would be _y - 4 = -2(x - 1)_. Converting it to slope-intercept form would give _y - 4 = -2x + 2_, and then _y = -2x + 6_. Each of these forms can be used to represent a linear equation, and different forms can be more useful depending on what information you have.

Graph characteristics, such as the slope and the intercepts, provide insights into the relationship between the variables. The slope tells you how much _y_ changes for a single change in _x_. Steeper lines have slopes with larger absolute values. A line that is closer to horizontal has a small absolute value slope, while a vertical line does not have a slope. The intercepts mark where the line crosses the axes. In real-world situations, the _x_ and _y_ intercepts often have significant meanings. For example, if we have an equation that models the total cost, _y_, of a purchase as related to the number of items purchased, _x_, the _y_-intercept will tell you the base cost before buying any item. The slope tells you the cost per item. Given a graph of a linear equation, you can extract the equation by determining the slope and _y_-intercept from the graph and using the slope-intercept formula.

Connecting graphical representations to algebraic equations allows you to interpret the mathematical relationship visually. When you have an algebraic equation, you can draw its line on the coordinate plane, which makes the relationship between variables much clearer. Likewise, given a graph, you can extract the equation by determining the slope and _y_-intercept. For example, if you have a line that goes through the points (0, 1) and (2, 5), the slope would be (5-1)/(2-0) = 2, and the _y_-intercept is 1, so the line equation would be _y = 2x + 1_. This interconvertibility between graphs and equations demonstrates their equivalence and allows us to translate between visual and symbolic representations of linear relationships.

Some common errors when graphing include plotting points incorrectly, which could stem from misreading coordinates or calculating values wrong, misinterpreting the signs of numbers (positive and negative), which leads to mis-plotting, drawing lines with incorrect slopes due to miscalculations or not using a ruler to ensure the lines are straight, and finally errors with the scale of the axes. These errors can lead to a distorted visual of the relationship between the variables. Always check your calculations and double check your points on your graphs. Using graph paper to plot the points more precisely will make graphing more accurate. When you identify a mistake, re-do the calculation or point immediately.

In summary, graphing linear equations allows us to visualize algebraic relationships. By learning to plot points on the coordinate plane, determine key points such as intercepts, calculate and interpret slope, and create accurate line graphs, students can improve their skills in algebra. Graphing linear equations is not just a mathematical exercise; it's a powerful tool to understand the relationships between variables, which has wide-ranging uses in many fields, including economics, physics, and engineering. By being able to visualize a linear equation, you are more prepared to handle real-world situations that are often represented using lines.

Crafting Linear Equations

Linear equations are a key tool in mathematics for describing the relationship between two variables and can be written in several different forms, each with its own specific uses and advantages. One of the most common forms is the slope-intercept form, written as $y = mx + b$, where m represents the slope of the line and b represents the y-intercept, the point at which the line crosses the y-axis. The slope, m, defines the steepness of the line and its direction, showing how much y changes for every one unit increase in x. For instance, in the equation $y = 2x + 3$, the slope is 2, meaning that for every one unit increase in x, y increases by two units. The y-intercept, b, provides the starting point of the line on the y-axis. In the same equation, $y = 2x + 3$, the y-intercept is 3, which means the line crosses the y-axis at the point (0, 3). The slope-intercept form is useful when you know the slope and y-intercept of a line, or when it's easiest to determine these values.

Another useful form for expressing a linear equation is the point-slope form, written as $y - y_1 = m(x - x_1)$, where m is the slope and (x_1, y_1)

represents a specific point on the line. This form is particularly handy when you know the slope of the line and a single point that the line passes through. The point-slope form makes it straightforward to write the equation of a line without needing to know the y-intercept directly. For example, if you know the line has a slope of 2 and passes through the point (1, 4), you can write the equation as $y - 4 = 2(x - 1)$. This form is a stepping stone to other forms, such as the slope-intercept form; you can simplify the equation to $y - 4 = 2x - 2$, and then add 4 to both sides to get $y = 2x + 2$, which is now in slope-intercept form. The point-slope form can be particularly helpful when dealing with real-world scenarios where you might know a rate of change and a particular value at a certain time.

The third primary way to express a linear equation is through standard form, which is written as $Ax + By = C$, where A, B, and C are constants. In this form, A and B cannot both be zero. Although it doesn't directly give the slope or y-intercept like other forms, it is useful in some situations. One advantage is that it easily allows for the use of integer coefficients, and it shows the relationship between the x and y variables in a balanced manner. To convert from standard form to slope-intercept form, you solve for y. For example, given the equation $2x + 3y = 6$, you would first subtract $2x$ from both sides to get $3y = -2x + 6$, then divide both sides by 3 to get $y = -\frac{2}{3}x + 2$, which is now in slope-intercept form. This conversion lets you see the slope and y-intercept of a line, making it easier to graph.

When writing a linear equation, it's important to choose the form that is most appropriate for the information given. If you know the slope and y-intercept, use the slope-intercept form. If you know the slope and a point, use the point-slope form. If you have an equation in standard form, you might want to rearrange it to one of the other forms for easier analysis, depending on what you need to do. It is useful to know how to convert between these different forms to make equations easier to work with. For instance, if you are given an equation in standard form but you need to graph it, converting it to slope-intercept form makes the graphing process smoother. The slope and y-intercept are clear, and this can help you graph the line.

A key skill in writing linear equations is finding missing information. For example, you might need to find the equation of a line given two points. In this case, you would start by finding the slope of the line. If you have two points, (x_1, y_1) and (x_2, y_2), the slope can be found using the formula $m = \frac{y_2 - y_1}{x_2 - x_1}$. Once you have the slope, you can then use either the point-slope form with either of the given points, or substitute the slope and one of the points into the slope-intercept form to find the y-intercept. For example, if you have the points (2, 3) and (4, 7), the slope would be $m = \frac{7 - 3}{4 - 2} = \frac{4}{2} = 2$. You can now use this slope and one of the points, say (2, 3), in the point-slope form: $y - 3 = 2(x - 2)$. This simplifies to $y - 3 = 2x - 4$, and finally to $y = 2x - 1$ in slope-intercept form.

Another situation where finding missing information is key is when you're given the slope of the line and one of the intercepts, either the x-intercept or the y-intercept. For example, let's say a line has a slope of -1 and crosses the y-axis at the point $(0, 5)$. This means the y-intercept is 5, and we already know the slope is -1. So, we can directly use the slope-intercept form to create the equation, which will be $y = -1x + 5$, or simply $y = -x + 5$. Alternatively, if you know the slope and the x-intercept, you would set $y = 0$ in the slope-intercept form, using the x-intercept as the point where the line crosses the x-axis. This will let you solve for the y-intercept, then write the full equation.

Verifying the accuracy of an equation is an important step in any problem-solving process. You can verify an equation by checking if the points on the line satisfy the equation. If a point is on the line, then the x and y values of that point should make the equation true when you substitute them into it. For example, if you have the equation $y = 2x + 1$, and you have a point (3, 7), substitute $x = 3$ and $y = 7$ into the equation: $7 = 2(3) + 1$, which simplifies to $7 = 6 + 1$, which is true. If the resulting statement is not true, you have likely made a mistake in either calculating the equation, or by testing the wrong point. Checking points on a line is one way to catch any errors and to make sure your equation is correct. You can always check your work multiple ways to verify its correctness.

Linear equations can be used to model a wide range of real-world scenarios. For instance, imagine a scenario where a taxi charges a flat fee of $5 and then $2 per mile. You can create a linear equation to model the total cost of a taxi ride. In this case, y could represent the total cost, x the number of miles, $2 the slope and $5 the y-intercept. So, the equation would be $y = 2x + 5$. If you travel 10 miles, you could find the cost by substituting $x = 10$ into the equation: $y = 2(10) + 5 = 25$, so the total cost would be $25. Linear equations help you to model many situations in real life.

Another example can be found in a business setting. Suppose a company has fixed costs of $1000 per month and variable costs of $5 per item produced. In this case, y can represent the total monthly cost, and x can be the number of items produced. The equation would then be $y = 5x + 1000$. Using this equation, the company can find its cost for any number of items produced, and can then use that information for planning. These examples show how linear equations provide a model to convert real-world scenarios into a useful math representation.

Understanding how to interpret the components of the linear equation is just as important as being able to create it. In the slope-intercept form, the slope always represents the rate of change, or how much the dependent variable changes with respect to the independent variable. For example, in the previous taxi fare example, the slope of 2 represents the rate at which the fare increases per mile driven. The y-intercept represents the base value or the starting value. In the taxi fare example, the y-intercept of 5 represents the base fee. In the business cost example, the 1000 represents the fixed costs, which are the costs that do not vary with the number of items produced. Knowing how to interpret the different elements of the equation will give you a deeper understanding of what the equation is telling you about the relationship it models.

Practice is essential when it comes to mastering linear equations. When writing these equations, start by thinking about the information you have. Do you have a slope and y-intercept? Do you have a slope and a point? Or do you have two points? The answer to this question will often point you towards the most useful form of the equation you need to use. For example, if you have two points (1, 5) and (3, 9), you must find the slope and then use that information to write the equation. First, find the slope using the formula: $m = \frac{9 - 5}{3 - 1} = \frac{4}{2} = 2$. Now, you can use either the point-slope form or the slope-intercept form to get the equation. Using the point-slope form with the point (1,5), you would write the equation as $y - 5 = 2(x - 1)$. You can also rewrite this to the more conventional $y = 2x + 3$ if you prefer. Practicing these steps can lead to improvements in your understanding of linear equations.

When faced with complex problems, breaking them down into smaller pieces can help. Identify the variables, the known information, and what you are trying to find. For example, if a car travels at 60 miles per hour, and you want to write an equation showing the distance it has traveled after a certain amount of time, you might let y be the distance and x be the time. Since the rate of change is a constant 60 miles per hour, the equation will have a slope of 60. In this example, the y-intercept is 0, as it would make sense that the car has traveled 0 miles if no time has passed. The equation representing this relationship would therefore be $y = 60x$. Always think about what you have and what you need in an equation.

Error analysis is an important part of developing equation skills. If an answer doesn't look right, review all the steps you took to get the equation. Start by double-checking the slope calculation. Did you subtract the y values in the correct order? Check all the points and calculations made and look for any potential problems. Then, check your substitutions and simplifications as well. If you've written the equation in point-slope form, check to make sure you plugged the correct values into the equation, or that your algebra is correct. With practice, you will start to identify the most common mistakes you make and can avoid repeating them. Error analysis can be a very useful learning tool.

Systems of Linear Equations Solved

Systems of linear equations, a collection of two or more linear equations with the same variables, can be solved using several methods, each offering unique advantages depending on the nature of the equations. One approach is the graphing method, where each equation is represented as a line on the coordinate plane. The solution to the system is the point, or points, where the lines intersect. This intersection point represents an ordered pair (x, y) that satisfies both equations simultaneously. When two lines intersect at a single point, the system has one unique solution; but when lines overlap perfectly, indicating they are essentially the same line, then the system has infinite solutions because every point on the line satisfies both equations. Alternatively, if two lines are parallel, they never intersect, and, therefore, the system has no solution. The graphing method provides a visual understanding of the solutions, making it a strong choice for introducing the topic or for situations where the relationships between variables should be made very apparent.

The substitution method is an algebraic approach that starts by solving one of the equations for one variable in terms of the other. This expression is then substituted into the other equation, effectively creating a new equation in one variable, which can then be solved directly. After obtaining the numerical value of one variable, it can then be plugged back into any equation containing both variables to solve the other unknown value. For example, take the system consisting of the equations $y = 2x + 3$ and $3x - y = 1$. In the first equation, we already have y isolated in terms of x. So, substitute $2x + 3$ for y in the second equation: $3x - (2x + 3) = 1$. Simplify this to get $3x - 2x - 3 = 1$ or $x - 3 = 1$. Add 3 to both sides to solve for x: $x = 4$. Once you have $x = 4$, plug it into the first equation: $y = 2(4) + 3 = 8 + 3 = 11$. So, the solution of the system is the point $(4, 11)$. The substitution method is efficient when one equation is easily solved for a variable, making it a smooth algebraic approach for solving for unknowns in a system.

The elimination method is particularly useful when the coefficients of one of the variables in the two equations are the same or opposite, or if they can be made to be through multiplication. In this method, we adjust the equations, usually by multiplying one or both equations by a constant, so that when we add or subtract the equations, one variable cancels out, leading to a single variable equation. For instance, consider the system $2x + 3y = 7$ and $4x - 3y = 5$. Notice that the coefficients of the y terms are opposites of each other. If we add the two equations together, the y terms will cancel out: $(2x + 3y) + (4x - 3y) = 7 + 5$, which simplifies to $6x = 12$. Now, we divide by 6 to solve for x: $x = 2$. Substituting $x = 2$ into the first equation, we get $2(2) + 3y = 7$, which simplifies to $4 + 3y = 7$. Subtract 4 from each side and then divide by 3 to find $y = 1$. Thus, the solution is $(2, 1)$. This method is very useful when the coefficients are convenient for canceling one variable and it can be very efficient. If the coefficients are not convenient, we can multiply one or both equations by a constant to make them suitable for the elimination process.

Selecting the most effective method for solving systems of equations often depends on the specifics of the equations. For example, if one equation is already solved for a variable, or can be easily solved, substitution might be the most direct method. If the coefficients of a variable are opposites or the same, elimination is a good choice. Graphing, while it is visually useful, is most effective when the solutions are integer values or where a visual representation is beneficial. It is important to consider all these options to find the most direct path to a solution. Sometimes, combining steps from different methods can lead to an efficient solution; for example, one might use substitution to reduce the number of variables and then apply elimination to further simplify the equations. Understanding the characteristics of a system of equations can help to choose the right method. A set of parallel lines indicates no solution, while a set of coincident lines signals infinite solutions.

A unique solution occurs when the lines intersect at a single point, meaning that the system has a single ordered pair that satisfies all of the equations. This point is located where both lines cross each other on a coordinate plane. This single point of intersection is the sole solution that makes both equations true at the same time. This is the most straightforward outcome of a system of equations, where each variable is assigned a particular value. When a system of equations has a unique solution, it suggests a specific relationship between the variables.

Infinite solutions, on the other hand, are found when the equations graph to the same line. This is called a coincident line situation. This indicates that the two equations are just different forms of the same equation. Because they are the same, every point on that line satisfies both equations. These systems present an interesting situation, as they show an unlimited number of solutions and illustrate how one equation may be converted into another via algebraic manipulation. If you find all of the variables canceling out of the problem and getting something like $0 = 0$, then there are infinite solutions.

A system of equations has no solution when the lines are parallel. Parallel lines never intersect on a coordinate plane and thus there is no point (x, y) that can satisfy both equations. If you are solving the problem and all of the variables cancel out, and you get a statement that is not true, such as $0 = 5$, then the system has no solution. The lack of intersection indicates that the equations are inconsistent and that there is no common solution. This lack of solution highlights the fact that not all sets of equations have a solution, and it demonstrates how equations must be consistent to have any solution.

Systems of equations have many uses in real-world problems. For instance, consider a situation where a business is trying to optimize its pricing. If a company sells both a standard version of a product and a premium version, and they know how many units of each they sold during a particular period, and they also know their total revenue, they can use a system of equations to calculate the price of each item. Another case might be the classic problem of mixing two different solutions, each with a different concentration, to achieve a particular concentration in the final solution. Or, consider calculating the speeds of two trains given their total travel distance and their relative speeds to one another. In all of these examples, variables can be used to represent the unknown quantities, and linear equations are used to represent the relationships between these variables. Solving these sets of equations will then give the needed values for the unknowns.

When interpreting solutions, it is vital to understand what each variable represents in the context of the problem. In the business example, the variable might represent the price of a good, and in a physics example, it might represent the speed of an object. The meaning of the solution should be understood in terms of the real-world situation. It is also important to check if the solution makes sense in the context of the problem. A negative speed, for example, might indicate an error in the equation formulation. Mathematical models are effective when they correspond with observable reality, so solutions must be validated to ensure that the equations were properly set up and solved, and they also must match what makes sense in the given scenario.

Systems of linear equations are a fundamental tool in a number of fields. In economics, they are used to analyze supply and demand, while in engineering they are used to calculate forces and analyze circuits. They form the basis of more advanced mathematical concepts and models. Being able to solve systems of equations is an essential skill for all mathematics students, and it is important in many other fields. A good grasp of these systems allows for modeling, analyzing, and solving all kinds of problems.

The best method for understanding the process is through guided exercises. Begin with simple systems that require the use of one method, and then advance to more complex systems. Encourage practice in all solution methods so the student learns how to choose the best method for a given situation. It can be useful to start with systems that have unique solutions and then advance to problems with infinite solutions, or no solutions, to have a good understanding of the full range of cases. It is also helpful to develop practice problems that force the student to compare methods, forcing a decision about the best way to go for a given situation.

Complex problem challenges push understanding of concepts to a higher level. Such problems might involve real-world scenarios with more equations, requiring careful analysis and use of more advanced algebraic manipulation. These challenges highlight the need to translate complicated situations into mathematical expressions. They require students to connect theory with practice, a process essential for the development of problem-solving skills. The ability to move between a situation and the equations used to solve it is a very useful tool. Error analysis is another important component of solving these more complex problems. If something does not make sense in the context of the problem, check the steps one at a time to identify any possible algebraic or logical errors.

Method comparison activities are useful because they teach the student that the equations often have a variety of solutions that can all be correct. These activities can be set up so that the student has to apply a variety of methods to the same problem. This will show that the solutions do not change as long as you perform all calculations accurately. Comparing methods allows the student to be more flexible in their approach, increasing their overall problem-solving capabilities. It will also allow the student to choose the most efficient path to a correct answer.

Linear Equations in Real-World Contexts

Linear equations serve as a powerful tool for understanding and solving problems found in everyday life, acting as a bridge between abstract mathematics and practical applications. Understanding how to translate real-world situations into mathematical equations, identify the patterns that suggest a linear relationship, and extract the needed mathematical information is vital for applying these concepts effectively. Consider a scenario in financial modeling, where businesses often use linear equations to represent the relationship between cost, revenue, and profit. For instance, if a company manufactures a product with a fixed cost of $500 per month plus a variable cost of $10 per unit, the total cost can be represented by the linear equation $C = 10x + 500$, where C is the total cost and x is the number of units produced. To understand the concept of break-even, which is the point where a company's total revenue equals its total cost, we would set the revenue equation equal to the cost equation and solve for the number of units. For example, if the company sells each unit for $25, the revenue equation would be $R = 25x$. Setting R equal to C, we have $25x = 10x + 500$, which simplifies to $15x = 500$, and $x = 33.33$. This indicates the business must sell approximately 34 units to break even. This shows how linear equations are important tools for businesses to understand their profitability and set financial goals.

In the context of investments, simple interest is a classic application of linear equations. If an investor deposits a principal amount (P) with an annual interest rate (r) for a time period of (t) years, the simple interest earned can be expressed as $I = Prt$, where I represents the interest earned. For example, if an investor deposits $1000 at a 5% interest rate for 3 years, the interest earned would be $I = 1000 \times 0.05 \times 3 = 150$, so the total amount after three years would be $1150. This model demonstrates a linear growth of investment over time.

Linear equations also play a role in scientific calculations. Distance-rate-time problems, which involve relating the distance traveled by an object, the rate at which it travels, and the time it takes, are expressed through the equation $d = rt$, where d is the distance, r is the rate (or speed), and t is the time. For instance, if a car travels at 60 miles per hour for 2 hours, the distance it covers would be $d = 60 \times 2 = 120$ miles. This can be used to find any of the three variables if the other two are known. Temperature conversions provide another use of linear equations in the sciences. The conversion between degrees Celsius (C) and degrees Fahrenheit (F) can be represented by the linear equation $F = \frac{9}{5}C + 32$. If the temperature is 20 degrees Celsius, its equivalent in Fahrenheit would be $F = \frac{9}{5}(20) + 32 = 68$ degrees. These types of equations are vital for accurate measurements and conversions across different scientific disciplines.

In social science contexts, linear equations can assist in modeling population changes and analyzing survey results. For example, if the population of a town is growing at a steady rate each year, that growth can be expressed using a linear equation. If a town starts with a population of 10,000 and grows by 500 people each year, its population (P) after (t) years can be modeled as $P = 500t + 10000$. This linear model allows for the projection of population sizes in the future. Linear equations also have applications in survey analysis. For example, if survey data reveals a linear correlation between the hours of study and test scores, the data can be modeled with a line equation to help understand this relationship. Consider also resource allocation problems, which often utilize linear relationships between available resources and the needs they meet. If there is a linear connection between resources applied and outcome produced, linear equations can be a useful tool for allocating these resources.

Solving real-world problems effectively requires a systematic approach that starts with identifying key variables in the situation, followed by creating a mathematical model, which can often be in the form of a linear equation. Interpreting the solutions within the context of the problem is important to see if it makes sense. Consider a multi-step word problem: Suppose a store sells two types of products, A and B. The store sells product A for $20 each, and product B for $30 each. If, in total, the store sold 50 products in a day and made a total of $1300, we could create a system of linear equations to find the number of each type of product they sold. Let 'x' represent the number of product A sold, and 'y' represent the number of product B sold. The equations would then be: $x + y = 50$ (total number of products) and $20x + 30y = 1300$ (total revenue). Using either substitution or elimination methods, the system can be solved to determine that the store sold 20 of product A and 30 of product B.

Real-world scenario analysis is another way to strengthen understanding of linear equations. Suppose a company plans to make a large purchase and is trying to decide between two options: purchase a new machine with fixed costs, but lower per unit costs, or purchase a used machine with higher per unit costs but lower initial costs. The company can model the cost of each choice with a linear equation, compare each case, and then make the best decision based on the number of units they expect to produce. This kind of analysis is very useful when modeling and making business decisions. Through the application of linear equations, we are able to turn these complex scenarios into solvable math problems.

One type of complex challenge involves looking for linear equations in diverse fields, like trying to use linear equations to model the rate at which a lake is filling up after a storm, or to calculate the costs of running a delivery service that has various fixed and per-mile costs. It's also possible to create a model of the relation between the number of hours a person studies and their exam results, or the amount of time a factory spends producing goods and the total profit they make.

Through practice with these complex types of problems, one can improve the understanding of how to translate situations into math equations, and how to interpret the meanings of the results. Through guided practice, students can also learn how to spot errors in their calculations, how to solve complex problems, and how to evaluate the solutions in the context of the problem they are trying to solve. This approach enhances not only the ability to calculate the right answers, but it also helps to form a greater understanding of the power of mathematics in everyday life.

Linear equations, therefore, serve as a universal language for understanding and addressing a wide range of real-world problems. From financial decisions to scientific analysis and social trends, the application of linear equations provides a structured approach to solving for the unknowns, making them an invaluable resource for both academics and professionals.

3

QUADRATIC EQUATIONS AND FUNCTIONS

Parabolas and Vertex Transformations

A parabola is a distinctive U-shaped curve that visually represents a quadratic function. It's characterized by several key features, including a vertex, an axis of symmetry, and a specific direction in which it opens. The vertex is the turning point of the curve, representing either the minimum or maximum value of the function. The axis of symmetry is a vertical line that passes through the vertex, dividing the parabola into two mirror-image halves. A parabola opens upwards if the coefficient of the squared term is positive, or downwards if it's negative. These basic components are used to understand the quadratic functions.

The vertex form of a quadratic equation, written as $y = a(x - h)^2 + k$, provides a clear method to identify these features and quickly sketch the graph of a parabola. In this form, 'a' determines the width and the direction of the opening of the parabola, while (h, k) represents the coordinates of the vertex. The value of 'a' also decides whether the parabola opens upwards (if 'a' is positive) or downwards (if 'a' is negative). A larger absolute value of 'a' creates a narrower parabola, while a smaller absolute value results in a wider one. This coefficient directly affects how stretched or compressed the basic parabolic shape will look on a graph.

The values of 'h' and 'k' in the vertex form dictate the horizontal and vertical shifts of the parabola from its basic position at the origin. Specifically, 'h' controls the horizontal translation. The graph moves to the right when 'h' is positive and to the left when 'h' is negative. Note that it's (x - h) in the equation, so a positive 'h' results in a shift to the right. The 'k' value is the vertical translation: a positive 'k' shifts the parabola upwards, and a negative 'k' shifts it downwards. Therefore, the vertex coordinates are (h, k).

To visualize these transformations, imagine the basic parabola $y = x^2$, which has its vertex at (0, 0). When we change the vertex form to $y = (x - 2)^2 + 3$, we shift the entire parabola 2 units to the right (because of the (x - 2) part) and 3 units up (because of the + 3). The new vertex is at (2, 3). Similarly, if we use $y = -2(x + 1)^2 - 4$, the graph opens downward (because of the negative sign in front of 2), is narrower than $y = x^2$ (because of the coefficient 2) and is shifted 1 unit to the left and 4 units down, with the vertex at (-1, -4). The combination of these changes affects both the appearance and position of the graph significantly.

Graphing a parabola using the vertex form is straightforward. First, identify the vertex from the values of 'h' and 'k' in the equation. Then, determine whether the parabola opens upward or downward based on the sign of 'a'. To get a more precise graph, you can find a few additional points by substituting different x-values into the equation and calculating the corresponding y-values. Plot these points, including the vertex, and then connect them to draw a smooth, U-shaped curve, keeping the symmetry in mind. For example, to graph $y = 2(x - 1)^2 + 2$, identify the vertex at (1, 2). Since 'a' is 2, it opens upward. Pick two points near the vertex, such as x = 0 and x = 2. When x = 0, $y = 2(0 - 1)^2 + 2 = 4$, so we have the point (0, 4). When x = 2, $y = 2(2 - 1)^2 + 2 = 4$, giving us (2, 4). Plot the points (0, 4), (1, 2), and (2, 4), and draw a smooth U-shape through them, reflecting across the line x=1 (the axis of symmetry). This method works for most quadratic equations presented in vertex form.

Let's walk through an example that involves various transformations. Consider the equation $y = -0.5(x + 3)^2 - 1$. The negative sign in front of 0.5 tells us that the

73

parabola opens downward. The coefficient of 0.5 makes the graph wider than the base parabola $y = x^2$. The $(x + 3)$ term indicates that the parabola shifts 3 units to the left from the origin. The term -1 moves the parabola 1 unit downward from the horizontal axis. Therefore, the vertex is at (-3, -1). The transformations result in a wide, downward-opening parabola with the vertex at (-3, -1). By plotting more points around the vertex, we can generate an accurate graph showing how every component of the equation has affected the resulting parabola.

Now it's time for some practice problems to reinforce these ideas. First, consider the equation $y = (x - 4)^2 + 1$. What is the vertex of this parabola, and in which direction does it open? The vertex is (4, 1), and since the coefficient 'a' is 1 (a positive number), the parabola opens upward. Sketch this graph using the steps previously discussed, and be sure to find a couple of points to either side of the vertex. Next, try this: $y = -2(x + 1)^2 - 3$. Identify the vertex and direction of opening here. The vertex is (-1, -3), and because of the negative coefficient, the parabola opens downward. This parabola is narrower than the basic parabola, $y=x^2$, and it is shifted to the left and downward. You should be able to picture these transformations just by understanding the changes made in vertex form, allowing you to easily sketch the graph.

Another practice question would be to analyze the equation $y = 3(x - 2)^2 - 5$. Identify the vertex, direction of opening, and whether it's narrower or wider than $y=x^2$. The vertex is (2, -5). The positive coefficient means the parabola opens upwards, and the 3 means it's narrower than $y = x^2$. Once you practice identifying these components of the equation, graphing them becomes a simpler task. Try sketching this parabola and observe the impacts of each term on the final graph. Can you see how the value of 'a' affects the parabola's width and direction? And how 'h' and 'k' change its position?

Another exercise could ask you to describe the transformations of the basic parabola $y=x^2$ to produce $y = -(1/2)(x+5)^2+3$. First, the negative coefficient means the parabola opens downwards. The 1/2 indicates that the parabola is wider than the basic one. The (x+5) term means that the parabola is shifted 5 units to the left of the origin. Finally, the +3 means that the parabola is shifted 3 units up. Putting this all together, you should see how a basic quadratic can be transformed.

Consider one more question. Suppose you are given a parabola that has a vertex of (3,-2) and goes through the point (4,-1). How could you write the equation for this parabola? We know the vertex, so we can plug h=3 and k=-2 into vertex form: $y = a(x - 3)^2 - 2$. Now we can use the point (4,-1) to solve for 'a'. Since the point (4,-1) falls on the graph, we know that when x=4, y=-1. Plugging this into our equation, we get $-1=a(4-3)^2 - 2$, which simplifies to -1 = a - 2. Solving for 'a' we get a=1, which gives us the final equation $y = (x - 3)^2 - 2$. You will have to apply these steps when you come across similar problems.

Understanding quadratic functions and how their vertex form reflects graphical changes is vital for using them to model real-world situations. For example, in physics, projectile motion can be described using quadratic equations, where the vertex of the parabola can represent the maximum height the object reaches. In business, the profit function may take the shape of a parabola, where the vertex represents the production or pricing level at which maximum profit is achieved. Understanding the graph gives you the key insights into real-world problems. Analyzing these examples reveals how essential a deep knowledge of parabolas and their transformations is to apply quadratic functions in different areas of study, and we will be using that to expand this study guide into more complex math.

Solving Quadratic Equations Masterfully

Solving quadratic equations involves finding the values of the variable that make the equation true, and this can be achieved through several methods, each with its own strengths and applications. Factoring, the first method, relies on rewriting the quadratic expression as a product of two linear expressions. This technique is effective when the quadratic expression can be easily factored. For example, consider the equation $x^2 + 5x + 6 = 0$. We seek two numbers that multiply to 6 and add to 5. These numbers are 2 and 3, so the factored form of the equation is $(x + 2)(x + 3) = 0$. Setting each factor to zero yields $x + 2 = 0$ or $x + 3 = 0$, giving the solutions $x = -2$ and $x = -3$. Factoring is most efficient with quadratics that have integer roots, but can be more difficult for those with non-integer or irrational roots.

The next technique, using the square root property, is applicable to quadratic equations in the form of $x^2 = c$ or $(x - h)^2 = c$, where 'c' is a constant. To solve $x^2 = 9$, take the square root of both sides, resulting in $x = \pm 3$, meaning the solutions are $x=3$ and $x=-3$. When using $(x - 2)^2 = 16$, taking the square root gives $x - 2 = \pm 4$. Then, adding 2 to both sides gives us $x = 2 \pm 4$ which means $x=6$ or $x=-2$. The key is to isolate the squared term and then apply the square root to find both positive and negative roots.

Completing the square, another valuable technique, transforms a quadratic equation into a perfect square trinomial, making it easier to solve. The steps involve ensuring the coefficient of the x^2 term is 1, moving the constant term to the right side of the equation, taking half of the coefficient of the x term, squaring it, and adding it to both sides of the equation. Consider the equation $x^2 + 6x + 5 = 0$. First, move the constant to the right side: $x^2 + 6x = -5$. Then, take half of 6 (which is 3), square it (which is 9), and add 9 to both sides, resulting in $x^2 + 6x + 9 = -5 + 9$, which simplifies to $(x + 3)^2 = 4$. Now we can use the square root property, giving $x + 3 = \pm 2$. Solving for x gives $x = -3 \pm 2$, which leads to the solutions $x = -1$ and $x = -5$. Completing the square is useful in deriving the quadratic formula and is important when rewriting a quadratic in the form $a(x - h)^2 + k$, called the vertex form of a quadratic, which we covered in the last section.

The quadratic formula is a universal method for solving any quadratic equation written in standard form as $ax^2 + bx + c = 0$. The formula is: $x = [-b \pm \sqrt{(b^2 - 4ac)}]$

/ 2a. In this formula, 'a', 'b', and 'c' are the coefficients from the standard form of the quadratic equation. Applying this to solve $2x^2 - 5x + 2 = 0$, we identify a = 2, b = -5, and c = 2. Substituting these values into the quadratic formula yields x = $[5 \pm \sqrt{((-5)^2 - 4(2)(2))}] / (2 * 2)$, which simplifies to x = $[5 \pm \sqrt{(25 - 16)}] / 4$, which is x = $[5 \pm \sqrt{9}] / 4$. So x = $[5 \pm 3] / 4$. This gives two solutions x = (5 + 3)/4 = 2 and x = (5 - 3)/4 = 1/2. The quadratic formula works for any quadratic equation, regardless of whether the roots are rational, irrational, or even complex.

The discriminant, which is the expression under the square root in the quadratic formula ($b^2 - 4ac$), provides important information about the nature of the roots of the quadratic equation. If the discriminant is positive, the equation has two different real roots. If it is zero, the equation has exactly one real root (which is considered a repeated root). If the discriminant is negative, the equation has two complex (non-real) roots. Looking back at our previous example, $2x^2 - 5x + 2 = 0$, the discriminant is $(-5)^2 - 4 \cdot 2 \cdot 2 = 25 - 16 = 9$. Because the discriminant is a positive 9, we know that our solutions will be two different real roots. When the discriminant is zero, we will have one real root. For example, if we solve $x^2 - 6x + 9 = 0$ using the quadratic equation, the discriminant is $(-6)^2 - 4 \cdot 1 \cdot 9 = 36 - 36 = 0$. In this case, you will get x=3 as your only solution. When the discriminant is negative, you will have non-real or complex roots. For example, if we solve $x^2 + x + 1 = 0$, the discriminant is $1^2 - 4 \cdot 1 \cdot 1 = 1 - 4 = -3$. This will give us non-real solutions since the square root of a negative number is non-real.

Let's go over some examples for practice. First, consider the quadratic equation $x^2 - 8x + 15 = 0$. We can try factoring this. We need two numbers that multiply to 15 and add to -8. Those numbers are -3 and -5. So, the factored form of the equation is (x - 3)(x - 5) = 0. Setting each factor to zero, we get x - 3 = 0 or x - 5 = 0, which gives solutions x = 3 and x = 5. Next consider the equation $2x^2 + 7x + 3 = 0$. We will need to use the quadratic formula here. With a = 2, b = 7, and c = 3, substituting into the formula gives x = $[-7 \pm \sqrt{(7^2 - 4 \cdot 2 \cdot 3)}] / (2 * 2)$, which simplifies to x = $[-7 \pm \sqrt{(49 - 24)}] / 4$ or x = $[-7 \pm \sqrt{25}] / 4$, which results in x = [-7

± 5] / 4. So, the two solutions are x = (-7 + 5)/4 = -1/2 and x = (-7 - 5)/4 = -3. Try solving x^2 - 10x + 25 = 0. This can be factored easily into (x-5)(x-5) = 0, which can also be written as $(x-5)^2=0$. Using the square root property we get x-5=0, giving us x=5.

Let's also solve x^2 + 4x - 12 = 0 using completing the square. We move the 12 over to get x^2 + 4x = 12. Half of four is 2, and 2 squared is 4. So, we get x^2 + 4x + 4 = 12+4, which results in $(x+2)^2=16$. Using the square root property, we get x+2 = ±4. This leads to x = -2 ±4, which means x=2 and x=-6. Finally, solve x^2 + 4x + 8=0, using the quadratic equation. Plugging in a=1, b=4, and c=8 to the equation gives x = [-4 ± √(4^2 - 4 1 8)] / (2 * 1), which simplifies to x = [-4 ± √(16 - 32)] / 2 or x = [-4 ± √-16] / 2. This will give us complex solutions since the discriminant is -16. You will have x = [-4 ± 4i]/2, which reduces to x = -2 ± 2i.

Choosing the right method depends on the specific nature of the quadratic equation. Factoring is quicker if the roots are integers, but not always applicable, and completing the square will often be more complex than other techniques. The square root property works when there is no 'bx' term in the equation. The quadratic formula is a reliable method that solves any quadratic equation but can sometimes be more time-consuming for simple quadratics. Understanding the discriminant helps determine the nature of the roots beforehand, saving time and guiding solution approaches.

Real-world problems often involve quadratic equations. For example, consider a ball thrown upward with an initial velocity. The height of the ball over time can be described by a quadratic equation, where the roots of the equation represent the times when the ball hits the ground. Or the area of a rectangle might be described by a quadratic equation, and solving the equation might determine the dimensions of the rectangle. Understanding how to solve these equations and interpret the results can lead to useful insights in many practical scenarios. The skills acquired solving quadratic equations will be used repeatedly in many areas of math, such as geometry, calculus, and physics, making these techniques a crucial part of any math education. Practice and understanding the properties of quadratic equations are vital for continued success in math.

Quadratic Real-World Problem Solving

Quadratic functions provide a powerful way to model many real-world scenarios, moving beyond simple linear relationships to describe curved paths and optimal points. The ability to solve quadratic equations becomes essential when applying them to practical situations.

One common area where quadratic functions shine is physics, particularly in the analysis of projectile motion. When an object is launched into the air, its path follows a parabolic arc, assuming we ignore air resistance. This is because gravity exerts a constant downward acceleration on the object, affecting its vertical motion while horizontal motion remains constant. A typical model for this situation, where height h is a function of time t, looks like: $h(t) = -1/2gt^2 + v_0t + h_0$, where g is the acceleration due to gravity (approximately 9.8 m/s^2 or 32 ft/s^2), v_0 is the initial vertical velocity, and h_0 is the initial height. This equation is a quadratic, revealing that the object's vertical position changes with time following a parabolic path. For example, consider a ball thrown upwards with an initial velocity of 20 m/s from a height of 2 meters. The equation governing its height over time would be $h(t) = -4.9t^2 + 20t + 2$. Using this, one can determine the time it takes to reach its maximum height, its maximum height, or the time when it will hit the ground. The vertex of the parabola, determined by completing the square or using the formula $t = -b/2a$, gives the time at the maximum height, and the y-value of the vertex gives the maximum height. To find when the ball will hit the ground, we would solve for when $h(t) = 0$ using the quadratic formula. These kinds of calculations can be applied to sports, such as determining the trajectory of a ball or the path of a projectile.

Engineering utilizes quadratic functions for design optimization. Engineers use this mathematical tool to model and analyze structures such as bridges and arches. For instance, the shape of an arch or a suspension cable can often be described using a parabola. This is used to determine structural integrity and load bearing capacities. In bridge building, knowing the maximum height and width of a parabolic arch is key to safely distribute weight and design support structures. The dimensions of parabolic reflectors, which are commonly used in satellite dishes and solar panels, are also determined using the properties of quadratic functions. For instance, a parabolic reflector is designed to focus incoming rays to a single point, the focus, which needs precision calculations. Area and volume calculations also incorporate quadratic functions, especially when dealing with shapes that are not simple rectangles or cubes. If you are trying to maximize the area of a rectangle with a limited amount of fencing, for example, the relationship between the length and width of the enclosure will be quadratic, as will the equation for the area. The vertex of the parabola will give the maximum area, which can determine the ideal dimensions.

Quadratic equations are also powerful in the realm of economics and financial modeling. Cost and revenue functions for businesses often involve quadratic terms. For example, total revenue might be described by a function such as $R(x) = -ax^2 + bx$, where x is the number of units sold. The inclusion of the x^2 term indicates that after a certain number of sales, the revenue will start to decrease as the demand will get saturated. This is a more realistic representation than a linear equation and might be due to price drops as you try to sell more goods. Cost functions also use quadratic terms, especially when considering economies or diseconomies of scale. An example of a cost function might look like $C(x) = ax^2 + bx + c$, where 'c' is the fixed cost, 'b' is the cost per unit, and the ax^2 term can indicate the increasing cost of producing more goods. The relationship between costs and revenues is crucial for determining break-even points, which are when the company makes no profit or loss. This can be modeled by a system of equations and solved algebraically, visually, or numerically. Break-even points can be found where the cost function intersects the revenue function. Companies often aim to optimize their profits, which means finding the production level where the difference between revenue and cost is greatest. The profit equation, $P(x) = R(x) - C(x)$, is often a quadratic function, and finding its maximum, which will be the vertex of the parabola when the x^2 term is negative, will determine the ideal number of units to maximize profits. In practical financial contexts, quadratic equations help in calculating the growth of investments when interest rates are not fixed, and they can also be used in risk management to understand potential losses associated with different investments.

When using quadratic equations to solve word problems, the initial step is to translate the language into mathematical expressions. Identify the key variables involved in the problem. Look for relationships between these variables that can be represented using quadratic functions. A projectile problem might contain details about the initial velocity and starting height, while an economic problem may give information about costs and revenues. After defining the variables, construct a quadratic equation to represent the situation. For example, if the problem involves an area, the equation might involve multiplication of variables representing dimensions, which often leads to a quadratic form. The next step is to select the appropriate method for solving the equation: factoring, the quadratic formula, using the square root property, or completing the square. The choice depends on the form of the equation and ease of use of each method. After obtaining the solutions, it is crucial to interpret the results in the context of the original problem, which means ensuring the answers make sense. For instance, a negative time value in a physics problem is not meaningful, and so you would ignore the negative answer. In an economic situation, you will need to round numbers up or down to the nearest whole unit if you cannot have a fraction of a unit. This kind of analysis allows for an understanding of the solutions to the problem and their implications in real-world situations.

To further grasp these concepts, consider the following diverse problem set. First, a rocket is launched vertically upward from the ground with an initial velocity of 100 meters per second. What is the maximum height the rocket will reach? In this case, we can use $h(t) = -4.9t^2 + 100t$, because we are starting at the ground where the initial height h_0 is zero. The vertex, where $t = -100 / (2 \setminus -4.9)$, gives the time the rocket reaches its maximum, and we can plug this value into our equation to find the height at this time. Next, suppose a farmer wants to fence off a rectangular field next to a river, using 600 meters of fencing material. The side adjacent to the river will not require fencing. What are the dimensions of the field that maximize the area? If we let x be the width and y be the length, we have $2x + y = 600$, so $y = 600-2x$. The area $A = x \setminus y = x(600-2x) = -2x^2 + 600x$. The vertex of this parabola, where $x = -600 / (2 \setminus -2)$, gives the value of x which maximizes the area. A final problem: A company's profit from selling x units is given by $P(x) = -0.1x^2 + 50x - 5000$. How many units should the company sell to maximize profit? Again, the vertex of the parabola, where $x = -50 / (2 \setminus -0.1)^*$, will give the value of x which maximizes profit, and that number of units should be produced and sold. Solving problems like these requires a mastery of skills such as setting up the equations, choosing the right solving method, and making sure your answers make sense within the context of the problem. This kind of critical thinking is invaluable in many areas of math and problem solving.

The versatility of quadratic functions as a mathematical tool is evident across diverse fields. Whether modeling physical motion, optimizing engineering designs, or analyzing financial models, quadratic equations can describe and predict behavior in complex real-world systems. This ability stems from the nature of these equations, that is they are a result of a variable being multiplied by itself, which leads to curves and optimums. The principles explained here are the base knowledge for many other areas of math. The methods covered here are essential for any student studying math, engineering, business, and many other topics.

Navigating Quadratic Inequalities

Quadratic inequalities, an extension of quadratic equations, introduce the concept of a range of solutions rather than just specific points, which adds another layer of complexity to working with quadratic functions. These inequalities involve comparing a quadratic expression to a value using symbols like

The fundamental difference between quadratic equations and inequalities is that equations have specific solutions where an expression equals a value, while inequalities have a range of solutions. When you see an inequality sign, it indicates that there are multiple numbers that can satisfy the condition. The symbol '≤' means "less than or equal to," while '≥' means "greater than or equal to." The symbols '

Solving quadratic inequalities typically follows a structured approach involving both algebraic manipulations and critical point identification. First, rewrite the inequality so that one side is zero; for instance, transform _$ax^2 + bx + c > d$_ into _$ax^2 + bx + (c - d) > 0$_. This sets the stage for finding the critical points, which are the x-values where the corresponding quadratic equation (_$ax^2 + bx + (c - d) = 0$_) equals zero, which are the x intercepts of the parabola. These critical points are the boundaries that separate the x-values that make the inequality true from those that make it false. These critical points can be found by using the methods we discussed earlier in the chapter, including factoring, the quadratic formula, or completing the square, based on the form of the quadratic. Once the critical points are found, they divide the number line into intervals. These intervals are regions where the inequality is either true or false. To determine if an interval is part of the solution, pick a number from within that interval and substitute it into the original inequality. If the inequality is true, the entire interval is part of the solution set; if false, it's excluded. Each interval must be checked, and all the intervals that satisfy the inequality condition comprise the overall solution.

Graphically, a quadratic inequality's solution set can be understood by relating the algebraic solution to the visual representation on a coordinate plane. To graph the inequality, begin by graphing the corresponding quadratic equation as a parabola. The parabola's shape and position are defined by the quadratic's coefficients, specifically the _a_, _b_, and _c_ values as explored earlier in the chapter. The critical points of the inequality are the x-intercepts of the parabola. If the inequality includes "equal to," then the parabola is drawn as a solid line; otherwise, it is drawn as a dotted line to indicate the boundary is not part of the solution. For a "greater than" inequality (_>_ or _≥_), shade the area above the parabola; for a "less than" inequality (_

When solving quadratic inequalities, several steps can ensure accuracy. Begin by making sure the inequality is in standard form, with zero on one side. This sets up the equation for the next step, which is to identify critical points. After finding the critical points, use them to divide the number line into intervals, and then use test points within each interval to see whether it satisfies the inequality condition. If the test point works, that region on the number line is a valid solution. When graphing, use the correct boundary line style—solid for inclusive inequalities and dotted for exclusive ones. Shade the appropriate region based on the direction of the inequality. A common mistake is to forget to flip the inequality sign when multiplying or dividing by a negative number in the solving stage. Another error is not considering all intervals when testing for solutions, which can result in missing parts of the overall solution. Also, make sure to check that the shaded portion of the graph corresponds to the solutions you found algebraically to make sure no errors have been made.

The application of these methods can be made clearer through specific examples. Consider the inequality _x^2 - 5x + 6 < 0_. To solve this, first factor the quadratic as (_x - 2_)(_x - 3_) < 0. The critical points are x=2 and x=3, these are the roots of the equivalent quadratic equation. These points divide the number line into the intervals _x

Next, solve the inequality $-2x^2 + 4x + 6 \geq 0$. First, divide the entire inequality by -2, remembering to flip the inequality sign, yielding $x^2 - 2x - 3 \leq 0$. Factoring provides the critical points: $(x-3)(x+1) \leq 0$. The critical points are $x=-1$ and $x=3$. Testing intervals for x

Let's solve the inequality $x^2 + 4x + 5 > 0$. If you try to use the quadratic formula on $x^2 + 4x + 5 = 0$, the discriminant is negative, so there are no real roots. In this case, there are no critical points. Because the a term (coefficient of x^2) is positive, the parabola opens upwards. Testing any value for x will prove that this inequality is always true. This means that for every x value on the number line, the parabola is always greater than zero, and so the solution set includes all real numbers. Graphically, this is shown with a parabola above the x-axis, and shading all the area above the parabola. If the a term was negative instead, the parabola would point downward, and the solution would be "no solutions."

Another example: solve the inequality $x^2 - 6x + 9 \geq 0$. Factoring gives us $(x-3)^2 \geq 0$, so we only have one critical point, x=3. Because a square is always positive or zero, this inequality is always true, and so the solution set includes all real numbers, just like in the previous example. The graph is a solid parabola, with its vertex on the x-axis at x=3, and shading the entire graph (every point is greater or equal to zero).

Quadratic inequalities are an important mathematical instrument used to describe a wide range of real-world scenarios, including physics, economics, and engineering. They help determine the conditions under which variables meet specific constraints, which can inform decision-making and analysis of real-world issues. For instance, understanding projectile motion often requires quadratic inequalities to find out the time range when a projectile is above or below a certain height. In business, inequalities might determine production levels where profits meet minimum targets. In engineering, this can be about creating designs that fit within specific performance boundaries. Mastering these concepts builds a strong base in algebraic thinking and problem-solving, essential for success in further math studies and practical applications. This ability to understand and analyze inequalities is not just about manipulating mathematical symbols; it is about building a solid framework for logical thinking and problem solving.

4

Exponential Functions and Properties

Exponential Growth Fundamentals

Exponential growth describes situations where a quantity increases at a rate proportional to its current value, leading to rapid acceleration over time, whereas linear growth involves a steady increase by the same amount in each time period. Linear growth can be visualized as a straight line on a graph, with the same increase between each point, in contrast, exponential growth forms a curve that rises slowly at first and then shoots up very rapidly. To illustrate, imagine two scenarios: a plant that grows two inches every week (linear growth) and a population of bacteria that doubles every hour (exponential growth). In the linear case, growth is constant, but with the bacteria, growth speeds up as the population increases.

Exponential growth happens when a quantity multiplies by a constant factor over equal intervals of time. Mathematically, we can represent this as $y = ab^x$ where y is the final quantity, a is the starting quantity, b is the growth factor, and x is the number of time periods. The growth factor is a key aspect; if it is greater than 1, the quantity increases, and if it is less than 1, the quantity decreases, which is called exponential decay. The value of b can be directly related to the percentage growth rate as well, with a growth of 5% being equivalent to a growth factor of 1.05 and similarly a decay of 5% being equivalent to a decay factor of 0.95.

Consider a real-world scenario like bacteria population growth. If you begin with one bacterium that doubles every hour, after one hour there will be two bacteria. After two hours, each of those two will double, resulting in four bacteria. After three hours, this process will repeat to produce eight bacteria. This results in a sequence of 1, 2, 4, 8... which grows faster with each new unit of time, demonstrating exponential growth. The equation for this scenario would be $y = 1 \cdot 2x$ where y is the number of bacteria and x^* is the number of hours.

Another practical illustration of exponential growth can be found in finance with compound interest. Unlike simple interest, where you only earn interest on the principal amount, compound interest earns interest on both the principal and any accrued interest. Suppose you invest $100 at an annual interest rate of 5%, compounded annually. After one year, you will have earned $5 in interest for a total of $105. In the second year, you will earn 5% interest on $105, which is $5.25. The new total after two years would therefore be $110.25. Each subsequent year, the interest will grow faster as the base amount keeps getting bigger. The formula $A = P(1 + r/n)nt$ describes this where A is the future value of the investment/loan, including interest, P is the principal investment amount (the initial deposit or loan amount), r is the annual interest rate (as a decimal), n is the number of times that interest is compounded per year, and t is the number of years the money is invested or borrowed for. This formula describes how the compounding effect can significantly increase the amount of money earned over time.

Conversely, exponential decay describes a situation where a quantity decreases by a constant percentage rate per unit time. A good example of exponential decay is radioactive decay. Radioactive substances decay as their unstable atoms release energy and transform into more stable forms. Each radioactive material has a specific half-life which is the time it takes for half of the material to decay. For example, if a radioactive material has a half-life of 10 days and a sample begins with 100 grams, after 10 days 50 grams will remain. In another 10 days, half of the 50 grams will decay, leaving 25 grams. The amount keeps decreasing at a slower rate, but it never reaches zero. In general, the function for exponential decay is $y = a(1 - r)x$ where y is the amount of the substance left after x periods, a is the starting amount, and r is the rate of decay.

To understand how exponential growth and decay rates are determined, it's important to recognize the significance of the growth/decay factor. The factor, denoted by b in $y = abx$, determines whether the function is growing or decreasing. If b is greater than 1, it indicates growth. The rate can be derived from this factor; for example, if b equals 1.07, it means the quantity is increasing by 7% each period. If b is less than 1 (but still greater than 0), this represents decay. A decay factor of 0.92, for example, means the quantity decreases by 8% each time period, where the decay rate is (1 - 0.92) * 100%.

Calculating the growth or decay factor from data involves comparing values at different time points. For example, if a population increases from 100 to 115 in one time period, the growth factor is 115/100, which equals 1.15. This factor translates to a growth rate of 15%. If the amount of a substance decreases from 80 to 60 in one period the decay factor is 60/80 which is 0.75 or a 25% decay rate per time period.

Graphically, exponential functions produce distinct curves. Exponential growth appears as a curve that starts low and rapidly increases, which becomes steeper as x increases. Exponential decay appears as a curve that starts high and steadily decreases towards the x-axis which approaches zero as x increases. The key differences from linear functions, which graph as straight lines, are the rate of change which is constant for linear functions, and constantly changing for exponential functions. These distinctions are critical for spotting the correct function that models the given situation.

To more easily grasp the contrasts between exponential and linear functions, consider a visual comparison. If you plot the functions of $y = 2x$ (a linear function) and y = 2^x (an exponential function) on the same graph, the linear function will be a straight line with a steady upward slope, whereas the exponential function will start slowly and then rise much quicker as x increases. This visual illustration clearly highlights how exponentially increasing quantities outpace linear ones over time and is essential in understanding the power and applications of exponential models. These models are essential in population dynamics, finance and science for projecting trends or changes.

Mastering Exponent Rules

Exponents provide a shorthand method for showing repeated multiplication, and understanding the rules that govern them allows for manipulating and simplifying expressions efficiently. When multiplying exponential terms that share the same base, you add the exponents. For example, xa multiplied by xb is equal to xa+b. To illustrate, consider 23 multiplied by 22. 23 means 2 2 2, which is 8, and 22 means 2 2, which is 4. Multiplying these results, we get 8 4 = 32. Using the rule, 23 multiplied by 22 is equivalent to 23+2 or 25, which also equals 32. This rule applies to any base, whether it's a number or a variable, provided the bases are the same. This method works because, when you have like bases, you are just counting how many times the base was a factor in the final product. For example, $\_x\_^2\_x\_^3$ is the same as _x__x__x__x__x_ or $\_x\_^5$. Be careful when you have different bases, for example $\_2\_^2\_3\_^2$. Because the bases are not the same, you can't simply add the exponents.

When dividing exponential terms with the same base, you subtract the exponents. The rule is: xa divided by xb equals xa-b. For instance, 35 divided by 32 can be solved by writing them out in long form. 35 is 3 \ 3 \ 3 \ 3 \ 3 or 243, and 32 is 3 \ 3 or 9. So the problem is 243 / 9 which equals 27. By the rule, 35 divided by 32 is 35-2 which is 33* and also equals 27. This rule is a direct result of the definition of division, which is the inverse operation of multiplication. When terms are written out, and the like terms are divided out from the numerator and the denominator, you are left with _a - b_ number of base terms. The result is always a base raised to the power of the difference of the exponents, and it is important to remember to subtract the exponent in the denominator from the exponent in the numerator. Note also that if the exponent in the denominator is bigger than in the numerator, the result will be a negative exponent, which will be explained later.

Another key rule is the power of a power rule. When you have an exponential term raised to another exponent, you multiply the exponents. For instance, $(x^a)^b$ is equal to x^{ab}. Imagine the expression $(4^2)^3$, which could be evaluated by first calculating 4^2 (which is 16), and then calculating 16^3 (which is 4096). Using the rule, you would just multiply the exponents 2 and 3 for a result of 4^6, which also results in 4096. A term like (x^2)^3 is equivalent to $(xx)(xx)(xx) = xxxxx^*x$, where the base 'x' is a factor six times or x^6. This method simplifies complex expressions and prevents common computational mistakes. When multiplying exponents, the power operation is being applied repeatedly, and you can think of the multiplication as counting the total number of times the base is a factor when there is an exponent outside of the parentheses.

A number of special cases of exponents include zero and negative exponents. Any non-zero number raised to the power of zero equals one. So, $x^0 = 1$, if x is not zero. For example, $5^0 = 1$, and $(-3)^0 = 1$. The zero exponent is derived from the rules of division. For example, x^3 divided by x^3 should equal x^{3-3} or x^0, but any number divided by itself is also always equal to 1, which provides a basis for the rule. It is important to note that 0^0 is not defined in standard algebra. A negative exponent represents the reciprocal of the base raised to the positive version of the exponent. For instance, x^{-a} is equal to $1 / x^a$. The negative exponent flips the position of the base term; if the base term is in the numerator, it moves to the denominator, and vice versa. Consider 2^{-3} which equals $1/2^3$, or 1/8. Similarly, $x^{-1} = 1/x$. The reciprocal rule is very important, and it is the same concept when the base term has a fractional exponent. For example, the exponent 1/2 is the same as taking the square root. The exponent 1/3 would be taking the cube root, and so on. So, when you have a rational exponent, the numerator becomes the normal power, and the denominator is the index of the root. For example, xa/b is the same as the b-th root of xa. Fractional exponents are useful in calculations and are very common when solving for exponential equations, as will be addressed in later sections of this book.

Common errors often occur with exponent rules. Students sometimes mistakenly add exponents during division, or subtract them when multiplying. Another mistake is to apply exponents to both the coefficients and base terms if there is more than one term in the base. For example, if the base is 2x and the exponent is 2 then the correct method is $(2x)^{\wedge}2 = 2^2 x^2 = 4x^2$. Students may also get confused when the base term is a fraction. For example, $(2/3)^2 = (2/3)*(2/3) = 4/9$ and the exponent applies to both the numerator and the denominator. Additionally, a negative exponent does not result in a negative number; it creates a fraction.

To solidify these concepts, consider the following examples:

1. Simplify a4a5: Applying the rule for multiplication of exponents with the same base, add the exponents: $a4+5 = a9*$.

2. Simplify (b6) / (b2): Applying the rule for dividing exponents with the same base, subtract the exponents: b6-2 = b4.

3. Simplify (c3)2: Using the power of a power rule, multiply the exponents: $c^{3\,2} = c6$.

4. Simplify 30: Any nonzero number to the power of zero is 1.

5. Simplify 4-2: This becomes 1 / 42 = 1 / 16.

6. Simplify x5/2: This becomes the square root of x5.

Here are some practice problems:

1. y2y7*

2. (z8) / (z3)

3. (m4)3

4. $(10)^0$

5. $(2)^{-4}$

6. n3/4

7. 2325 / 22*

8. $(32)3 / 3^4$

9. (x2y3)2*

10. a2 / a-3

11. $(4^*x^3)^2$

12. $(3/5)^3$

13. (x1/2)4

14. $(8x^6)/(2x^2)$

15. $(2^*x^{-2})^3$

16. $(4/x^{-2})$

By understanding and applying these exponent rules, one can simplify complex mathematical problems into more manageable forms. The ability to manipulate exponential expressions is a necessary skill for solving equations and understanding how to express large and small numbers in a compact way. Mastery of these rules builds a strong algebraic foundation and prepares students to tackle more complex mathematical concepts. This topic is also very important for any problems that use logarithmic functions, which will be addressed in the next chapter. Remember that practice is the best way to solidify your understanding. Work out the above practice problems and continually work through additional problems to reinforce what you have learned.

Solving Exponential Equations

Solving exponential equations requires different strategies depending on the structure of the equation, and it builds upon the rules for exponents that were previously discussed. These equations involve variables in the exponent and necessitate isolating that variable. One approach to solve exponential equations is to first attempt to make both sides of the equation have the same base. When the bases are the same, you can set the exponents equal to each other and solve for the variable. For instance, consider the equation $2^x = 8$. Here, 8 can be written as 2^3, so the equation becomes $2^x = 2^3$. Since the bases are the same, we can set the exponents equal to each other, resulting in $x = 3$. This is a straightforward method, but it only applies when the terms can be easily converted to the same base. Consider another example: $3^{(2x+1)} = 81$. Here, 81 can be written as 3^4, making the equation $3^{(2x+1)} = 3^4$. Equating the exponents gives $2x + 1 = 4$, which leads to $2x = 3$, and finally $x = 3/2$. When both sides of the equation cannot be easily converted to the same base, you will need to apply logarithmic transformations.

Logarithms provide a method for isolating a variable within an exponent when you have bases that are not the same. A logarithm is the inverse operation of exponentiation, and it allows you to bring the exponent down to base level where it can be more easily isolated. The logarithmic equation logb(x) = y can be read as "the logarithm of x to the base b equals y" and is equivalent to the exponential equation by = x. The most common logarithms are base 10, called the common logarithm, which is written as log(x), and base e (approximately 2.71828), called the natural logarithm, written as ln(x). In many cases any base will work to solve an equation, but the base 10 logarithm, which is usually on a calculator, is often the easiest to use. Consider the equation $5^x = 125$. Using the method of making bases the same, 125 is 5^3, so you know that x=3, but you can also use a logarithm to solve it. Taking the log of both sides gives $\log(5^x) = \log(125)$. A property of logarithms states that $\log(a^b) = b\log(a)$, which allows the exponent to be brought down: $x\log(5) = \log(125)$. To solve for x, divide both sides by log(5) so that x = log(125) / log(5). Using a calculator, log(125) is approximately 2.097, and log(5) is approximately 0.699, and 2.097/0.699 is 3, which is the correct answer. The base of the logarithm does not change the final answer, and this rule holds true for logarithms of any base. In fact, you can even use the base 5 logarithm, which would directly result in x = log5(125) = 3. When solving using logarithms, it is usually best practice to not round off intermediate values during a calculation so that you can get the most accurate final result. You can also use logarithms to solve more difficult problems. For instance, consider the equation $7^{(x+2)} = 13^x$. Here, taking the logarithm of both sides results in $\log(7^{(x+2)}) = \log(13^x)$. Applying the logarithm rule, we have (x+2) log(7) = x log(13). Distributing log(7) on the left side results in x log(7) + 2 log(7) = x log(13). Rearranging the equation gives x log(13) - x log(7) = 2 log(7). Factoring out the x results in x(log(13) - log(7)) = -2 log(7). Finally, dividing both sides by (log(13) - log(7)) results in x = -2 log(7) / (log(13) - log(7)). Using a calculator, this is x = -2 * 0.845 / (1.114 - 0.845), or x = -1.69 / 0.269, so x ≈ -6.29. Logarithms allow you to solve exponential equations even when the bases are different, and they are a powerful tool to understand exponential relationships. Logarithms can also be used for exponential decay problems. For example, the decay of a radioactive material is described by $A(t) = A_0 e^{(-kt)}$ where A(t) is the amount of

95

material remaining at time t, A0 is the initial amount of the material, and k is the decay constant. Using this equation, you could solve for k given the half life of a material, which would be when A(t) is half the value of A0. You would find that the natural log is best used to isolate the variable, as it is the base e logarithm.

Another technique for solving exponential equations is using graphical methods. This involves plotting the equation as a function on a coordinate plane and looking for the intersection of the graphed functions. When you plot functions, the x axis represents the variable and the y axis represents the function value, and it is easy to see how changing the x value can change the y value. Consider the equation $2^{\wedge}(x) = 8$. You can graph the function $y = 2^{\wedge}(x)$, as well as the function $y = 8$. Where the two lines intersect is the solution to the equation $2^{\wedge}(x) = 8$, and the x value at the intersection is the solution (x = 3). Similarly, for the equation $2^{\wedge}(x) = 5$, the intersection of $y = 2^{\wedge}(x)$ and $y = 5$ will be the solution. Here you will get an intersection that is not a whole number; you will need to use a calculator to approximate that the answer is near 2.32. The graphical method is particularly useful when you have more complex equations or where you can't find an exact algebraic solution. A graph can also show how many solutions there might be. For example, some equations may have more than one solution, which is apparent when the lines intersect in multiple spots. In general, you will only have a single solution for equations of the form $y = a^{\wedge}(x)$ when you are intersecting a horizontal line. If you have a more complicated equation such as $(x+1) = 2^{\wedge}(x)$, you may not find an easy algebraic solution. Plotting the two functions using a calculator is easy, and it can be easily found that the solution is near 2. A graph can also show that there are multiple solutions, which could be easy to overlook if you were just using algebraic techniques.

When solving exponential equations, you should follow certain rules to help reduce errors. First, always check your answer by substituting it back into the original equation. If the left and right sides of the equation are not equal, you know there was an error in your calculation. Always make sure that you are using the correct base for your logarithms, and remember that when there is no base specified, it is typically base 10. It is also a common error to not distribute negative signs through the parenthesis, so pay particular attention to detail. It is important to be consistent with rounding and to avoid premature rounding, to get the most accurate answer. Also remember that the rules for exponents apply to all variables and coefficients. Finally, you should be familiar with using different methods to solve the problem, since some techniques may be more efficient for certain types of problems.

To solidify these concepts, consider the following examples:

1. Solve for x: $4^{\wedge}(x) = 64$. Here you can make the base the same by noting that 64 is equal to $4^{\wedge}3$. So the equation becomes $4^{\wedge}(x) = 4^{\wedge}3$, so x = 3.

2. Solve for x: $3^{\wedge}(2x - 1) = 27$. You can make the base the same by noting that 27 is equal to $3^{\wedge}3$. So the equation becomes $3^{\wedge}(2x - 1) = 3^{\wedge}3$. Equating the exponents, 2x - 1 = 3, and so 2x = 4, and finally x= 2.

3. Solve for x: $5^{\wedge}(x) = 20$. The bases can't be made the same easily, so you must use logarithms. Take the log of both sides, $\log(5^{\wedge}x) = \log(20)$. Apply the rule to bring down the exponent: $x*\log(5) = \log(20)$. Divide both sides by $\log(5)$ to get $x = \log(20)/\log(5)$. Using a calculator this is approximately x = 1.301 / 0.699, or x ≈ 1.86.

4. Solve for x: $2^{\wedge}(x+1) = 3^{\wedge}(x-2)$. Start by taking the log of both sides: $\log(2^{\wedge}(x+1)) = \log(3^{\wedge}(x-2))$. Use the logarithm rule to bring the exponents down: $(x+1)\log(2) = (x-2)\log(3)$. Distribute $\log(2)$ on the left, and $\log(3)$ on the right: $x\log(2) + \log(2) = x\log(3) - 2\log(3)$. Then move terms with x to the left, and terms without x to the right: $x\log(2) - x\log(3) = -2\log(3) - \log(2)$. Then factor out the x: $x(\log(2)-\log(3)) = -2\log(3) - \log(2)$. Then divide both sides by $(\log(2)-\log(3))$ to solve for x: $x = (-2\log(3) - \log(2)) / (\log(2) - \log(3))$. Using a calculator, we get x

= (-2*0.477 - 0.301) / (0.301 - 0.477) which simplifies to x = -1.255 / -0.176 or x ≈ 7.13.

Here are some practice problems:

1. $2^x = 32$

2. $7^x = 49$

3. $3^{2x} = 81$

4. $4^{x+1} = 64$

5. $2^x = 10$

6. $5^{2x} = 100$

7. $6^{x+1} = 2^{x-1}$

8. $3^{2x+1} = 4^x$

9. $2^{x^2} = 16$

10. $3^x = 5^{x-1}$

11. $10^x = 1/100$

12. $(1/2)^x = 8$

13. $4^{2x} - 5 * 4^x + 4 = 0$ (Hint: let $y = 4^x$)

14. $3^x + 3^{x+1} = 108$

15. Use a graph to find the approximate solution to $2^x = x + 2$

16. Use a graph to find the approximate solution to $e^x = 4 - x$.

17. A bacterial culture doubles every hour. If there are initially 100 bacteria, how long will it take for the population to reach 10,000?

18. A radioactive substance decays such that half of it disappears in 10 years. What percentage of the substance will remain after 30 years?

19. The value of a car depreciates at 15% per year. If you bought the car for $30,000, when will the car be worth less than $10,000?

20. You invest $1000 in an account that earns 5% interest compounded annually. When will your investment double?

By working through these problems, and continuing to practice solving exponential equations, one can better understand the relationships between exponentiation, logarithms, and the graphs of these functions. This strong grasp of fundamental mathematical tools enables you to confidently apply them to various fields and to prepare for more complex mathematical problems.

Linear vs Exponential Dynamics

Linear and exponential functions represent two fundamental types of mathematical relationships, each with distinct characteristics and applications. Understanding the differences between them is key for grasping many mathematical and real-world phenomena. Linear functions exhibit a constant rate of change, whereas exponential functions display a rate of change that accelerates over time.

Mathematically, a linear function can be expressed in the form $y = mx + b$, where m represents the slope, or the constant rate of change, and b is the y-intercept, the point where the line crosses the y-axis. The slope indicates the steepness of the line and the direction of its rise or fall. For example, if a car travels at a constant speed of 60 miles per hour, the distance it covers over time can be modeled by a linear function. In this example, speed is the slope, and if you assume the car starts at zero distance then the y intercept is zero. This means each hour, the car travels an additional 60 miles, a consistent amount. A key feature of linear equations is the consistent addition of the same amount per each change of the independent variable.

In contrast, an exponential function is represented by the form y = ab^x, where a is the initial value, b is the base (which determines the rate of growth or decay), and x is the exponent. In an exponential function, the rate of change is not constant; it is proportional to the current value of the function. This means as the value of x increases, the function grows (or decays) at an ever-increasing pace. For instance, consider a population of bacteria that doubles every hour. The number of bacteria will not increase by the same amount each hour, it will double, and each hour there will be more bacteria than the previous, with a larger increase each time.

Visually, linear functions are always represented by straight lines on a graph. The constant slope is evident, with the line rising or falling at a consistent rate. The y-intercept is where the line crosses the y-axis at x=0, and the x-intercept is where the line crosses the x-axis when y=0. If the slope is positive, the line moves up from left to right, and if the slope is negative the line falls from left to right. This simple representation makes it easy to quickly see the rate of change and predict where values might lie. You can also plot a linear function from two points by finding the slope between those points and then using the slope-intercept equation.

Exponential functions, on the other hand, are represented by curves. Exponential growth curves start relatively flat, then curve upwards sharply, showing the rapid increase as x grows. Exponential decay curves start relatively high, and then curve down rapidly as x grows, approaching the x axis but never quite reaching it. The initial value, represented by a, is the point at which the curve intersects the y-axis, with a value at x=0. The base of an exponential function b is important for understanding how quickly it is growing (or decaying). If b is greater than 1, the graph grows, if b is less than 1 the graph decays. The larger the value of b, the more quickly the graph grows, and conversely if b is less than 1, then the lower it is the faster the graph decays.

The rates of change in linear and exponential functions are a major point of difference. Linear functions have a constant rate of change, meaning that for each unit increase in x, the value of y changes by the same amount. This makes it easy to predict what value y will be at any value of x as you simply add the same slope for each unit. Exponential functions, however, have a rate of change that increases (or decreases) as the function's value changes. Each increase in x causes a rate of change in y that is greater than the previous change.

Real-world applications of linear functions are numerous and can be seen in many different fields. One example is in calculating earnings when you are paid an hourly wage. The amount you earn is directly proportional to the number of hours you work, which is described by a linear function. You could take any number of hours worked, multiply by the hourly wage, and you know exactly how much money you earned. Another example is the distance a car covers moving at a constant speed over time, as discussed earlier. These examples illustrate that anytime there is a direct proportion between variables, that relationship is described by a linear function.

Exponential functions are used in cases where the rate of change is proportional to the value itself. Examples include population growth, compound interest, and radioactive decay. In population growth, the larger the population, the more individuals there are to reproduce, leading to a rate of change proportional to the population's current value. In compound interest, the interest you earn each period is added to the principal which increases future interest earned, and is another example of a quantity increasing in a way that is proportional to its current value. Radioactive decay occurs when an unstable isotope gives off particles which turns it into a stable isotope. The amount of an unstable isotope that decays each period depends on the number of unstable isotopes there are, such that at the start, a large amount decays, and as time passes and less isotope remains, less of it decays each period.

Predictive modeling with linear and exponential functions is used in different ways, because of the properties of these functions. With linear functions, it is easy to project outcomes because there is a constant rate of change. If you know a car travels 60 miles per hour you can accurately predict where it will be in 10 hours with no error. This is because linear functions always proceed along the same straight path. Exponential functions are not so simple. The initial change may not seem large, but the rate of change will get larger and larger as x grows. This makes it more difficult to predict exactly what will happen in the distant future, but the models are good for general trends and for knowing what might happen if a trend continues without any outside changes. For instance, if a population doubles each hour, you may think it will be small in a few hours, but in a very short amount of time the population will be so large that it will be difficult to manage.

To identify if a scenario is best represented by linear or exponential growth, consider how the variable is changing. If the amount of change is constant, then it's linear. If the rate of change is proportional to the current value, then it's exponential. For example, a plant growing two centimeters each week is a linear relationship, but a plant growing by 10% in height each week is exponential. Thinking about these relationships will help to understand the many real world examples.

The long-term behavior of these functions also differs greatly. Linear functions continue at a steady rate, never accelerating. Exponential functions start with a modest increase or decrease and then increase or decrease more and more dramatically as x gets larger. This is why it is important to consider these functions carefully when planning for the future, since an initial small amount can balloon in size very quickly if it is increasing exponentially.

Applying mathematical reasoning to real-world scenarios involves translating real world trends to a mathematical function that can then be used for prediction. This skill involves carefully looking at different scenarios and determining what the trend is that can be described by a mathematical function. When you understand these tools, you will be able to predict trends in many different settings. For instance, you could easily calculate how much you would earn at a job or you could predict how many bacteria there will be in a culture with exponential growth.

In conclusion, linear and exponential functions are essential mathematical tools for understanding and modeling the world around us. Linear functions model constant change, while exponential functions model growth or decay that is proportional to current values. Understanding these differences will help in identifying patterns in nature and using mathematics to solve real world problems.

5

POLYNOMIALS AND FACTORING

Polynomial Arithmetic Mastery

Polynomials, fundamental building blocks in algebra, are expressions consisting of variables and coefficients, combined using addition, subtraction, and multiplication, where exponents on the variables are non-negative integers. Each polynomial contains terms, which are individual parts separated by addition or subtraction signs. Terms consist of a coefficient (a numerical value) and a variable raised to a power, called its degree. For instance, in the polynomial $3x^2 + 2x - 5$, the terms are $3x^2$, $2x$, and -5; the coefficients are 3, 2, and -5, respectively, and the degrees of the terms are 2, 1, and 0 (since -5 is equivalent to $-5x^0$). The degree of the polynomial itself is determined by the highest degree of any term within it, which is 2 in the previous example.

To add polynomials, you combine like terms, that is, terms that have the same variable raised to the same power. For example, to add $(4x^3 + 2x^2 - x + 7)$ and $(x^3 - 5x^2 + 3x - 2)$, you group the terms with the same degree: $(4x^3 + x^3) + (2x^2 - 5x^2) + (-x + 3x) + (7 - 2)$. Combining the coefficients of these like terms, you obtain $5x^3 - 3x^2 + 2x + 5$. A common error when adding polynomials is incorrectly combining terms with different exponents, such as adding $2x$ and $3x^2$, which are not like terms. To avoid this, it is helpful to write each polynomial in descending order of exponents, to visually align like terms.

Subtracting polynomials requires distributing the negative sign to each term in the polynomial that is being subtracted. For example, to subtract $(2x^2 + 3x - 4)$ from $(5x^2 - x + 6)$, first rewrite the problem as $(5x^2 - x + 6) - (2x^2 + 3x - 4)$. Then distribute the negative sign, which changes the operation to $(5x^2 - x + 6) + (-2x^2 - 3x + 4)$. Now you can combine like terms as you would with addition: $(5x^2 - 2x^2) + (-x - 3x) + (6 + 4)$, resulting in $3x^2 - 4x + 10$. A typical error when subtracting is not distributing the negative sign to all terms within the parentheses, and this can lead to incorrect solutions. It is helpful to write out each term with its correct sign, to reduce the risk of mistakes.

Multiplying polynomials, on the other hand, requires more systematic distribution techniques. When multiplying two binomials (polynomials with two terms each), the FOIL method can be used, which is an acronym that represents the order in which to multiply the terms: First, Outer, Inner, Last. For instance, to multiply $(x + 2)$ and $(x - 3)$, first multiply the First terms, which gives x x = x^2. Then multiply the Outer terms, which results in x -3 = -3x. Next, multiply the Inner terms, giving 2 x = 2x. Lastly, multiply the Last terms, which gives 2 -3 = -6. Combining all the results leads to $x^2 - 3x + 2x - 6$, and then simplifying by combining like terms, you obtain $x^2 - x - 6$.

For multiplication involving polynomials with more than two terms, the expanded multiplication approach must be used, where each term in one polynomial is multiplied by each term in the other polynomial. For example, consider $(2x + 3)(x^2 - 4x + 5)$. Start by multiplying 2x by each term in the second polynomial: $2x(x^2) = 2x^3$, $2x(-4x) = -8x^2$, and $2x(5) = 10x$. Then multiply 3 by each term: $3(x^2) = 3x^2$, $3(-4x) = -12x$, and $3(5) = 15$. Combining all these products gives $2x^3 - 8x^2 + 10x + 3x^2 - 12x + 15$. Combining like terms simplifies this to $2x^3 - 5x^2 - 2x + 15$. When multiplying polynomials, it's essential to keep track of terms and their corresponding exponents, and some learners will find it helpful to use a grid or a table to organize the calculations.

Dividing polynomials can be more intricate. One method, polynomial long division, is similar to long division with numbers. For example, to divide $(x^2 + 3x + 2)$ by $(x + 1)$, set up the problem as a long division problem with $x + 1$ as the divisor and $x^2 + 3x + 2$ as the dividend. First, divide the first term of the dividend (x^2) by the first term of the divisor (x), which gives x. Write x above the x term in the dividend, and multiply the divisor $(x + 1)$ by x, obtaining $x^2 + x$. Subtract this result from the first two terms of the dividend $(x^2 + 3x)$. This yields $(x^2 + 3x) - (x^2 + x) = 2x$. Bring down the next term of the dividend (2) to obtain $2x + 2$. Divide 2x by x to get 2, and write 2 above the constant term in the dividend. Multiply the divisor $(x + 1)$ by 2 and obtain $2x + 2$. Subtract this result from the remaining dividend, which equals $(2x + 2) - (2x + 2) = 0$. The quotient is therefore $x + 2$. A remainder of zero means that $(x + 1)$ divides evenly into $(x^2 + 3x + 2)$, and the result is x+2.

Synthetic division is a faster method, that can be used when dividing by a linear divisor of the form $(x - a)$, where "a" is a constant. To divide $(x^3 - 6x^2 + 11x - 6)$ by $(x - 2)$ using synthetic division, write down the coefficients of the dividend: 1, -6, 11, and -6, and write the constant term of the divisor, which is 2. Bring down the first coefficient, 1. Multiply this by the divisor's constant, 2, to get 2, and write this below the next coefficient (-6). Add -6 and 2, which results in -4. Multiply -4 by 2 to get -8 and write it below 11. Add 11 and -8 to get 3. Multiply 3 by 2 to get 6 and write it below -6. Add -6 and 6 to get 0. The resulting numbers are 1, -4, 3, and 0. The 0 is the remainder and the other numbers represent the coefficients of the quotient, so the result is $x^2 - 4x + 3$. This method is particularly useful for factoring polynomials and determining zeros.

When dividing polynomials, you might encounter a remainder, which represents the part that does not divide evenly into the dividend. For example, dividing $x^2 + 2x + 5$ by $x + 1$ yields a quotient of $x + 1$ and a remainder of 4. This is expressed as $x + 1 + (4 / (x + 1))$. Recognizing that a zero remainder indicates perfect divisibility is important for identifying factors of polynomials. The interpretation of the remainder is also helpful for understanding function behaviors.

The ability to add, subtract, multiply, and divide polynomials has applications in many areas. In engineering, polynomial functions describe curves and shapes in various designs. In physics, they model the motion of objects. In business and finance, polynomial functions are used to analyze cost, revenue, and profit patterns. Furthermore, mastering polynomial operations provides a foundation for further study of algebraic concepts, such as rational functions, calculus, and more advanced mathematical and scientific courses. These operations are not abstract exercises, they are concrete skills needed for a range of real world problem solving. By understanding these operations, students develop an ability to see the world around them and solve real problems.

Advanced Factoring Strategies

Factoring is a fundamental skill in algebra, and a variety of strategies are essential for simplifying complex expressions and solving equations, and the following sections detail the most useful methods. First, identifying the greatest common factor (GCF) is the initial step in many factoring problems. The GCF is the largest factor that divides all terms of a polynomial evenly. To find the GCF, first identify the numerical GCF of all coefficients, and then identify the variable factors common to all terms, selecting the smallest exponent. For example, consider the polynomial $12x^3 + 18x^2 - 24x$. The numerical GCF of 12, 18, and 24 is 6, and the variable factor common to all terms is x, since the smallest exponent of x is 1. The GCF of the entire expression is therefore 6x. Factoring this out of each term of the polynomial, we obtain $6x(2x^2 + 3x - 4)$. Here, each term inside the parentheses is the result of dividing each corresponding term in the original polynomial by 6x: $(12x^3/6x = 2x^2)$, $(18x^2/6x = 3x)$, and $(-24x/6x = -4)$.

When factoring, it is important to first check if the polynomial has a GCF that can be factored out as the initial step. Failure to do this will often make further factoring very difficult. Another important step for students is to double check their work by redistributing the factored term back into the expression. If the result of the redistribution is not the original polynomial, it indicates an error in the factoring process. For instance, with the above example, multiplying out $6x(2x^2 + 3x - 4)$ will verify the correctness of the GCF process. It produces $12x^3 + 18x^2 - 24x$, which is identical to the original expression, confirming the accuracy of the factoring.

The next common factoring method is recognizing and applying the difference of squares pattern. A difference of squares occurs when two perfect squares are subtracted from each other, expressed as $a^2 - b^2$. This type of polynomial is easily factorable into the form $(a + b)(a - b)$. For example, consider the expression $x^2 - 9$. Here, x^2 is the square of x, and 9 is the square of 3, so the expression fits the difference of squares pattern, where a = x and b = 3. Therefore, the factorization of $x^2 - 9$ is $(x + 3)(x - 3)$. Another example is $4x^2 - 25$. Here, $4x^2$ is the square of 2x, and 25 is the square of 5, so the factorization is $(2x + 5)(2x - 5)$. Problems become more difficult when variables are in the squared terms, such as $16x^2 - 49y^2$, the square root of $16x^2$ is 4x and the square root of $49y^2$ is 7y, so the factored expression is $(4x+7y)(4x-7y)$. When factoring a difference of squares, students should make sure that the expression is in the proper form, with one term subtracted from another and both terms being perfect squares. An easy mistake is attempting to factor the sum of two perfect squares, such as $x^2 + 4$ which is not factorable with real numbers.

Trinomial factoring involves factoring quadratic expressions that have three terms, which usually take the form $ax^2 + bx + c$. There are several methods for factoring trinomials. One common method is the AC method, which involves multiplying the coefficient of x^2 (a) by the constant term (c), which equals ac. Then, find two numbers whose product is ac and whose sum is b, the coefficient of the x term. Suppose the trinomial is $2x^2 + 7x + 3$. Here, a = 2, b = 7, and c = 3. So, ac = 2 3 = 6. We need to find two numbers that multiply to 6 and add to 7. These numbers are 6 and 1, since 6 1 = 6 and 6 + 1 = 7. Next, rewrite the middle term (7x) as the sum of 6x and 1x, giving us $2x^2 + 6x + x + 3$. Then, factor by grouping. From the first two terms, factor out 2x, leaving 2x(x + 3). From the last two terms, factor out 1, leaving 1(x + 3). The expression now reads 2x(x + 3) + 1(x + 3). Since (x + 3) is common to both terms, factor it out to get (x + 3)(2x + 1). This is the factored form of the original trinomial.

Another method for trinomial factoring uses trial and error. This method begins by creating two sets of parentheses and filling in the terms whose product is the first term of the trinomial, and then finding terms whose products are the last term of the trinomial. The cross products must add to the middle term of the trinomial. For example, consider factoring $x^2 + 5x + 6$. First set up two sets of parentheses, ( )( ). Since the first term of the trinomial is x^2, we know the first term of each parenthesis is x, so we get (x)(x). Next, look for two numbers whose product is the constant term, 6, and whose sum is the coefficient of the x term, 5. Those two numbers are 3 and 2. Placing these into the parentheses gives us (x + 3)(x + 2). If the signs of the constant and middle terms are not positive, then the numbers that should be placed into the parenthesis will need to be negative or of opposing signs, depending on the signs of the trinomial terms. For instance, with $x^2 - 5x + 6$, the product is positive (6) and the middle term is negative (-5), therefore, both terms must be negative when filling the parentheses and the factored form is (x-2)(x-3).

When applying the AC method and grouping, common mistakes include incorrectly multiplying the 'a' and 'c' coefficients, misidentifying the two numbers whose sum is 'b' and product is 'ac', and improper factoring by grouping in the last step. Some students also may struggle with sign changes and may incorrectly choose positive factors when they should be negative factors or vice versa. With the trial and error method, common errors include selecting incorrect numbers that do not equal the constant or whose cross products do not equal the middle term. Practice with a range of trinomials, including those where the coefficient of x^2 is not one, is essential for mastering this skill.

Special factoring patterns, such as perfect square trinomials, simplify the factoring process. A perfect square trinomial is in the form $a^2 + 2ab + b^2$ or $a^2 - 2ab + b^2$, where the factored form is $(a + b)^2$ or $(a - b)^2$, respectively. For example, consider $x^2 + 6x + 9$. This is a perfect square trinomial because x^2 is the square of x, 9 is the square of 3, and 6x is twice the product of x and 3. Thus, the factored form is $(x + 3)^2$. Another example is $4x^2 - 12x + 9$. Here, $4x^2$ is the square of 2x, 9 is the square of 3, and 12x is twice the product of 2x and 3. Therefore, the factored form is $(2x - 3)^2$. Recognizing these patterns can simplify factoring problems considerably.

The sum and difference of cubes are two additional factoring patterns. The sum of cubes follows the pattern $a^3 + b^3$ which factors to $(a + b)(a^2 - ab + b^2)$, while the difference of cubes follows the pattern $a^3 - b^3$ which factors to $(a - b)(a^2 + ab + b^2)$. For example, consider factoring $x^3 + 8$. Here, x^3 is the cube of x and 8 is the cube of 2. Therefore, $x^3 + 8$ factors to $(x + 2)(x^2 - 2x + 4)$. Note that there is no further factoring that can be done with $x^2 - 2x + 4$. For factoring a difference of cubes, consider $x^3 - 27$. Here, x^3 is the cube of x, and 27 is the cube of 3. Therefore, $x^3 - 27$ factors to $(x - 3)(x^2 + 3x + 9)$, and again, there is no further factoring to be done with the second factor. A common mistake is to confuse the sum or difference of cubes with the sum or difference of squares, which have very different factored forms. Students must remember which pattern to apply.

In addition to recognizing these special patterns, some expressions might require multiple factoring techniques. For example, consider the polynomial $2x^3 - 8x$. First, the GCF, 2x, should be factored out, resulting in $2x(x^2 - 4)$. The resulting expression inside the parentheses, $x^2 - 4$, is a difference of squares, which then factors to $(x+2)(x-2)$, and so the fully factored polynomial is $2x(x + 2)(x - 2)$. A more difficult example is $3x^4 - 48$. Factoring out the GCF first gives $3(x^4 - 16)$, and recognizing $x^4 - 16$ as a difference of squares, the factored expression becomes $3(x^2 + 4)(x^2 - 4)$. Factoring again the difference of squares gives the fully factored polynomial as $3(x^2 + 4)(x + 2)(x - 2)$. For complex factoring scenarios, looking for a GCF first, and then applying any difference of squares or other factoring patterns, leads to proper factorization.

Diagnostic tips can help students recognize which factoring method to apply in different scenarios. First, always look for a GCF. Then, consider whether a binomial is a difference of squares or a sum or difference of cubes. If a trinomial is a perfect square, use that factoring pattern. If the trinomial does not fit the perfect square pattern, then use the AC method or trial and error techniques. Complex factoring scenarios often require multiple steps using combinations of the above strategies. Therefore, recognizing the forms of the factoring methods and the techniques for using each strategy will improve performance. A thorough practice will allow students to effectively apply these factoring methods in their algebra problems.

Polynomial Equation Solutions

Polynomial equations, which are equations where polynomials are set equal to zero, can be solved using a variety of methods, each suited to different types of equations. A foundational method for solving polynomial equations is using the zero-product property. This property states that if the product of two or more factors is zero, then at least one of the factors must be zero. For instance, if $(x - 2)(x + 3) = 0$, then either $x - 2 = 0$ or $x + 3 = 0$, which leads to solutions $x = 2$ and $x = -3$. The zero-product property allows us to break down a more complicated polynomial into smaller, easier to solve, linear equations. However, the polynomial must be in factored form in order to use this property, so factoring is often necessary for solving polynomial equations, and this is why the factoring skills in the previous section are so important.

Factoring is an essential first step in using the zero product property. By transforming a polynomial into a product of simpler factors, we can use the zero-product property to solve for the roots of the equation. For example, the quadratic equation $x^2 - 5x + 6 = 0$ can be factored into $(x - 2)(x - 3) = 0$. Applying the zero product property, this implies either $x - 2 = 0$ or $x - 3 = 0$, which gives solutions $x = 2$ and $x = 3$. Likewise, for a higher degree polynomial such as $x^3 - 2x^2 - 15x = 0$, the first step should be factoring out a common factor, which results in $x(x^2 - 2x - 15) = 0$. The expression in the parenthesis can be further factored into $(x-5)(x+3)$, resulting in $x(x-5)(x+3) = 0$. Using the zero product property, we now have $x=0$, $x-5 = 0$, and $x+3=0$, resulting in solutions of $x=0$, $x=5$, and $x=-3$. Factoring complex polynomials often requires using a combination of GCF extraction, special factoring patterns such as difference of squares or trinomial factoring. Students need to know how to recognize all of these factoring methods in order to apply the zero product property to solve polynomial equations.

The solutions to a polynomial equation are referred to as roots, and they represent the values of the variable that make the equation true. These roots can be real numbers, and as will be seen later, some can be imaginary or complex. For a polynomial equation, the number of roots, or solutions, is at most equal to the highest degree of the polynomial. For example, a cubic equation (highest power of 3) has a maximum of three roots, although these roots may not all be distinct. Real number roots correspond to x-intercepts on the graph of the polynomial, which will be described below. Each root signifies a specific point where the polynomial equation equals zero, which is of great use when understanding polynomial behavior.

Another way to find the solutions to a polynomial equation is through graphical approaches, particularly when factoring is difficult or not possible. Graphing calculators are invaluable tools that make it easy to visualize and analyze polynomial functions, allowing users to see the graphs and approximate the roots of polynomial equations. To solve a polynomial equation like $x^2 - 4x + 2 = 0$

graphically, you first input the corresponding function y = x² - 4x + 2 into the calculator. The calculator will display the graph of the parabola and show where the curve intersects the x-axis. These intersection points, or x-intercepts, indicate where y equals zero, therefore, these points correspond to the solutions of the equation x² - 4x + 2 = 0.

The x-intercepts of a polynomial graph represent the real number solutions of the corresponding polynomial equation, so by identifying where the graph crosses the x-axis, you identify the roots of the equation. Each x-intercept corresponds to a real root of the polynomial equation. For example, if the graph of a polynomial function crosses the x-axis at x = -1, x = 2, and x = 4, then the roots of the polynomial equation are -1, 2, and 4. In the previous example, x² - 4x + 2 = 0, the graph would cross the x axis at approximately x=0.59 and x=3.41, and this gives us the solutions for the roots of that polynomial. The number of x-intercepts is also directly linked to the number of real solutions of the equation. A polynomial of degree 'n' can have up to 'n' real number roots, and thus up to 'n' x-intercepts. It may have less than 'n' real roots, however, it will never have more.

There is a very direct relationship between the graph of a polynomial function and the solutions of the corresponding polynomial equation, and the graph is a visualization of the behavior of the polynomial and its roots. The shape of the graph, the number of x-intercepts, and the direction of the curve are tied to the solutions and the coefficients of the equation. For instance, a positive coefficient for the highest degree term in a quadratic polynomial means that its parabolic graph opens upwards, while a negative one opens downwards. The graph allows for not just finding solutions, but understanding the behavior of the polynomial, which is essential for deeper analysis.

In addition to factoring and graphing methods, algebraic strategies are essential when solving more complex polynomial equations, or when more precision is needed. These methods focus on using properties of equality to isolate the variable step-by-step and they involve a process of careful manipulation of the equation to find the unknown values. The goal of this method is to isolate the variable on one side of the equation, in order to clearly identify its value. This method is applicable across various polynomial types, from linear to more difficult expressions.

Solving polynomial equations algebraically often requires multi-step processes and a very methodical approach to maintain accuracy and to achieve the solutions. This may involve applying the distributive property, combining like terms, factoring, or using inverse operations. For example, the equation $2x^3 + 4x^2 - 10x = 0$ is a multi-step problem. The first step in solving this is to factor out the greatest common factor, $2x$, giving $2x(x^2 + 2x - 5) = 0$. This gives an initial root of $x=0$. If you try to factor the expression in the parentheses, you will not be able to find integer roots. In this case you need to rely on the quadratic formula to find the roots to $x^2 + 2x - 5 = 0$. The quadratic formula will give you the following solutions: $x=(-2+\sqrt{24})/2$, which equals approximately 1.45, and $x=(-2-\sqrt{24})/2$, which equals approximately -3.45. Therefore the solutions to $2x^3 + 4x^2 - 10x = 0$ are $x=0$, $x = 1.45$ and $x = -3.45$. This example highlights the importance of factoring as much as possible, and the need for using other algebraic methods when needed.

It is important to verify your solutions by substituting them back into the original equation, to ensure accuracy and correctness. By substituting the found values back into the initial polynomial equation, you check if the equation remains balanced and true. For example, given the equation $x^2 - 5x + 6 = 0$, after factoring and using the zero product property, you found solutions $x=2$ and $x=3$. When you substitute $x=2$ into the original equation, you will have $(2)^2 - 5(2) + 6 = 4 -10 + 6 = 0$. Therefore $x=2$ is verified to be a solution. When you substitute $x=3$ into the original equation, you will have $(3)^2 - 5(3) + 6 = 9 -15 + 6 = 0$, therefore $x=3$ is also verified as a correct solution. This verification process not only checks accuracy but also reinforces an understanding of how solutions make equations balanced and true.

For complex polynomials with multiple terms, rational or irrational roots, there are additional advanced strategies that can be used. Multi-step problems need more intricate strategies, like the example given above. These strategies may involve simplifying, rearranging, and using multiple algebraic steps to arrive at the solutions. Another challenging aspect for polynomial equations is dealing with rational roots, which are roots that can be expressed as a ratio of two integers. There are methods and theorems that can help find rational roots (such as the rational root theorem), and these can help guide your search for solutions. Irrational roots, on the other hand, cannot be expressed as a simple ratio and usually require methods such as the quadratic formula, or graphical solutions.

When facing a difficult polynomial equation, it's important to choose a suitable method. For some polynomials, factoring may be possible and will give exact solutions, while other polynomials may require the use of the quadratic formula. Other times, a graphical solution may be the only choice, or may be necessary to get approximate solutions. It is also a good practice to first attempt to find simple, integer roots using the factoring techniques already covered. Students must develop a strategic approach, considering the specific polynomial and deciding on the best method to solve for the roots.

Polynomial equations are not merely abstract algebraic concepts; they have practical significance in various real-world scenarios, and here are several real-world applications to consider. For instance, quadratic equations are used in physics to describe the trajectory of projectiles, helping to calculate the height or distance that an object travels through the air, given the initial velocity and launch angle. Engineers use polynomial equations to design bridges, buildings, and other structures, making sure that they can safely bear loads and withstand stresses. Financial analysts use polynomial equations to model the growth and decline of investments, and to understand changes in the market. Understanding and solving polynomial equations allows for a greater understanding of the world around you.

To solidify understanding, a wide variety of practice problems are essential for building the necessary skills for solving polynomials. Practice should include simple linear and quadratic equations, complex factored expressions, and higher order polynomials with integer, rational, and irrational roots. By combining factoring, graphing, and algebraic methods, students can master problem-solving and successfully tackle challenging polynomial equation problems. The goal is to have students apply these various solution methods, as this will reinforce their skills and help prepare them for the EOC exam and more advanced math courses.

6

RADICAL AND RATIONAL EXPRESSIONS

Mastering Square and Cube Root Simplification

Radical expressions, often involving square roots and cube roots, require specific simplification techniques to be fully understood. A square root of a number is a value that, when multiplied by itself, equals the original number; for instance, the square root of 9 is 3 because 3 multiplied by 3 equals 9. A cube root, on the other hand, is a value that, when multiplied by itself three times, gives the original number; for example, the cube root of 8 is 2, since 2 x 2 x 2 = 8. These operations are inverses of squaring and cubing, respectively, and serve as fundamental building blocks for algebraic manipulations.

Identifying perfect squares and perfect cubes is crucial for efficient simplification of radical expressions. Perfect squares are numbers that result from squaring an integer, such as 1, 4, 9, 16, 25, and so on. Knowing these numbers makes it quicker to spot opportunities for simplification under a square root. Similarly, perfect cubes are numbers obtained by cubing an integer, for example, 1, 8, 27, 64, 125, etc. Recognition of these perfect cube values is essential for working with cube roots. The basic properties of radicals, including the multiplication property $(\sqrt{ab} = \sqrt{a} * \sqrt{b})$ and the division property $(\sqrt{a/b} = \sqrt{a} / \sqrt{b})$, are powerful tools for breaking down complex expressions into simpler forms.

The process of simplifying radicals starts with breaking down the radicand (the number under the radical symbol) into its prime factors. This step is vital for finding perfect square or perfect cube factors hidden within the number. For example, to simplify $\sqrt{72}$, one might begin by factoring 72 into 2 x 36, then noting 36 is a perfect square. This allows the expression to be rewritten as $\sqrt{(2 \cdot 36)}$ or $\sqrt{2} \cdot \sqrt{36}$. Then the square root of 36 is simplified into 6, resulting in the expression $6\sqrt{2}$. The goal is to pull out the largest possible perfect square factor, leaving the smallest possible integer in the radicand. Similar strategies can be applied to cube roots where you look for perfect cube factors, extracting them from under the cube root symbol.

When simplifying mixed radicals (radicals that have a coefficient), the coefficients must be considered alongside the simplification of the radicand. Take for instance, the expression $3\sqrt{50}$. First, find the largest perfect square factor of 50 which is 25, rewriting the radical as $3\sqrt{(25 \cdot 2)}$, then $3 \cdot \sqrt{25} \cdot \sqrt{2}$. The square root of 25 is simplified to 5, which is then multiplied by the coefficient of 3 to yield $15\sqrt{2}$. Simplifying mixed radicals needs careful tracking of the coefficients. Mistakes can be easily made during the calculations if the coefficients are not considered. This procedure is applicable to cube roots, where you must look for perfect cube factors. For example in the expression $2\sqrt[3]{24}$, you look for the largest perfect cube factor of 24, which is 8. Rewriting the expression as $2\sqrt[3]{(8 \cdot 3)}$ or $2\sqrt[3]{8}\sqrt[3]{3}$ which simplifies to $2 \cdot 2\sqrt[3]{3}$ or $4\sqrt[3]{3}$.

To test the understanding of the principles, a series of problems should be practiced. The level of difficulty will increase, starting with simple problems like $\sqrt{16}$, where students should immediately recognize the perfect square and simplify to 4. Moving to $\sqrt{48}$, students should factor it to $\sqrt{16} \cdot \sqrt{3}$ and simplify to $4\sqrt{3}$. Similarly, for cube roots, practice could start with $\sqrt[3]{27}$, simplifying to 3 and move to more complex problems like $\sqrt[3]{54}$, breaking it down to $\sqrt[3]{27} \cdot \sqrt[3]{2}$ and getting $3\sqrt[3]{2}$. Mixed radicals with coefficients can be introduced next. For example, $2\sqrt{75}$ would require factoring 75 to $25 \cdot 3$, then simplifying to $2 \cdot \sqrt{25} \cdot \sqrt{3}$, which further simplifies to 2

5 $\sqrt{3}$ and resulting in $10\sqrt{3}$. For $3\sqrt[3]{16}$, one would factor 16 to 8 2, simplifying to 3 $\sqrt[3]{8}$ $\sqrt[3]{2}$, which then becomes 3 2 $\sqrt[3]{2}$, resulting in $6\sqrt[3]{2}$. These exercises ensure students can recognize when a radical can be simplified, can apply simplification steps, and can verify the result.

More challenging problems could involve multiple steps and larger numbers to test comprehension. For example, simplifying $\sqrt{192}$ would require students to break it down, maybe to $\sqrt{64}\sqrt{3}$ or $\sqrt{16}$ $\sqrt{12}$, and then simplify it to $8\sqrt{3}$ or $4\sqrt{12}$, requiring another round of simplification to get to the answer $8\sqrt{3}$. These problems require a more comprehensive understanding of prime factorization and radical properties. Another challenging example might be $5\sqrt{243}$, which would require factoring 243 to 81 3, resulting in 5 $\sqrt{81}$ $\sqrt{3}$ = 5 9 $\sqrt{3}$, simplifying to $45\sqrt{3}$. Likewise for cube roots, students might need to simplify $2\sqrt[3]{250}$. They should break down 250 into 1252, resulting in 2 $\sqrt[3]{125}$ $\sqrt[3]{2}$, and get 2 5 $\sqrt[3]{2}$ simplifying to $10\sqrt[3]{2}$.

Radical simplification is not merely a mechanical process; it also involves understanding when a radical expression cannot be simplified any further. For example, a radical like $\sqrt{15}$ cannot be simplified because 15 has no perfect square factors other than 1. Similarly, $\sqrt[3]{10}$ cannot be simplified. These examples underscore the importance of prime factorization and recognizing perfect squares and perfect cubes. Understanding this aspect will aid in preventing incorrect or incomplete simplification.

These challenging problems help students develop a deeper mastery of radical simplification and prepares them for a comprehensive understanding of algebraic principles for the next chapter and their upcoming EOC exam. The goal is not just to arrive at the correct answer, but to understand the methods of simplifying and applying the simplification techniques correctly. When dealing with more challenging examples, students can benefit from a methodical step-by-step approach that includes: first, factoring the radicand into its prime factorization; second, identifying the perfect square and perfect cube factors; third, extracting the perfect factors from the radical expression; and lastly, simplifying by combining coefficients. This approach promotes accuracy and comprehension, and prevents confusion.

In summary, this chapter on radical simplification requires a comprehensive approach involving understanding the definitions of square and cube roots, recognizing perfect squares and cubes, and applying the properties of radicals to simplify expressions. It involves a step-by-step strategy for simplifying, starting with prime factorization and then extracting perfect factors. Practice problems will aid in learning, building from simple to complex, and also addressing mixed radicals which have coefficients. Also included is the ability to determine when a radical cannot be simplified. A strong understanding of these principles is key for mastering algebra and for tackling more complicated mathematical problems. The skills gained are essential, not just for the Florida Algebra 1 EOC exam, but also for more advanced mathematics courses, where manipulating and simplifying radical expressions is a common task.

Rationalizing Denominators Effectively

Rationalizing denominators is a key technique in algebra that involves removing radical expressions from the bottom part of a fraction, transforming it into an equivalent form that is easier to work with. This practice stems from the preference in mathematics for having rational numbers, numbers that can be expressed as a ratio of two integers, in the denominator of fractions. An irrational expression involves numbers that cannot be expressed as a simple fraction of two integers, and these often include radical expressions. When an irrational number, such as $\sqrt{2}$, is in the denominator, it makes comparison and further mathematical operations more complicated, hence the need to rationalize. The goal of rationalizing is to convert the denominator into a rational number without altering the overall value of the fraction. This step simplifies expressions and makes them more suitable for additional algebraic operations.

For basic denominators containing a single radical term, the process is straightforward. For example, consider a fraction like $1/\sqrt{2}$. The goal is to remove the square root from the denominator. This is accomplished by multiplying both the numerator and the denominator of the fraction by the radical in the denominator. In this instance, both numerator and denominator are multiplied by $\sqrt{2}$. This yields $(1 \backslash \sqrt{2}) / (\sqrt{2} \backslash \sqrt{2})$. The numerator simplifies to $\sqrt{2}$, and the denominator, $\sqrt{2} \backslash \sqrt{2}$, becomes 2, because the square root of a number multiplied by itself equals the number. The result is $\sqrt{2}/2$. By multiplying by the radical, the denominator is transformed from an irrational number to the rational number 2, without changing the fraction's overall value. This method of multiplying by the radical in the denominator can be used for any single-term radical in a denominator. For instance, the fraction $3/\sqrt{5}$ would be rationalized by multiplying the numerator and denominator by $\sqrt{5}$, resulting in $(3 \backslash \sqrt{5}) / (\sqrt{5} \backslash \sqrt{5})$, which simplifies to $3\sqrt{5}/5$. Note that the same technique applies to fractions that involve cube roots, but in this case, you multiply by the cube root such that the denominator would become a rational number after simplification. For instance, in the fraction $1/\sqrt[3]{2}$, you would multiply the numerator and denominator by $\sqrt[3]{(2^2)}$ resulting in $\sqrt[3]{4} / \sqrt[3]{8}$ or $\sqrt[3]{4}/2$.

More complex denominators may involve more than just a single radical term, often involving a sum or a difference that includes a radical. For instance, take the fraction $1/(1+\sqrt{2})$. In these situations, simply multiplying by the radical in the denominator will not work. The key is to use the conjugate. The conjugate of an expression of the form $(a + \sqrt{b})$ is $(a - \sqrt{b})$, and vice versa. The key thing about conjugates is that their product is always a rational number. Multiplying $(a + \sqrt{b})$ by $(a - \sqrt{b})$ will result in $a^2 - b$. For example, the conjugate of $1 + \sqrt{2}$ is $1 - \sqrt{2}$. To rationalize the denominator of $1/(1+\sqrt{2})$, multiply both the numerator and denominator by the conjugate, $1 - \sqrt{2}$. The resulting expression will be $(1 \backslash (1 -$

√2)) / ((1 + √2) \ (1 - √2)). In the numerator, this becomes 1 - √2. The denominator, when multiplied, is (1 \ 1) + (1 \ -√2) + (√2 \ 1) + (√2 \ -√2), which simplifies to 1 - √2 + √2 - 2, and then further to 1 - 2, or -1. Therefore, the rationalized form of 1/(1+√2) is (1 - √2) / -1 or, simplified to -1 + √2. When you need to rationalize fractions that involve conjugates, careful attention should be paid to the signs and the distribution. For a fraction like 2/(√3 - √5), multiply by the conjugate √3 + √5. Multiplying this out, in the numerator will be 2(√3 + √5) = 2√3 + 2√5, and in the denominator (√3 - √5) * (√3 + √5) is equal to 3 - 5, or -2. So the fraction is equal to (2√3 + 2√5) / -2, and after simplifying it will be -√3 - √5.

To identify when rationalization is required, look for radicals in the denominator of fractions. This is usually the first step in simplifying an expression involving radicals. For example, in the expression (3 + √2) / (√5), the radical in the denominator indicates that the denominator must be rationalized. To do this, one multiplies both the numerator and the denominator by √5, resulting in ((3+√2) \ √5) / (√5 \ √5), which is (3√5 + √10) / 5. Checking this fraction, there is no radical left in the denominator. In more complex expressions, such as (1 - √3) / (2 + √3), the presence of the radical term in the denominator requires multiplication by the conjugate, 2 - √3. This yields ((1 - √3) \ (2 - √3)) / ((2 + √3) \ (2 - √3)), which simplifies to (2 - √3 - 2√3 + 3) / (4 - 3), and becomes (5 - 3√3) / 1, or just 5 - 3√3. After performing these operations, it is important to check for simplification. In many cases, both numerator and denominator may have common factors that can be reduced. In the example with the conjugate, (2√3 + 2√5)/-2 becomes -√3 -√5 by dividing both terms by 2.

To check the result, re-evaluate the final expression. If possible, compare the new form to the original using a calculator to ensure both results are the same. For example, one can input the original fraction 1/(1+√2) and the result -1 + √2 into a calculator and confirm they both are equivalent, and verify that the denominator of the result does not have radicals.

Here are practice problems to help solidify understanding:

1. Rationalize 2/√3. Here you multiply by √3/√3, resulting in 2√3/3.

2. Simplify 5/(2√2). Multiply by √2/√2. Result is 5√2/4.

3. Rationalize 1/(√5 - 2). You multiply by the conjugate (√5 + 2). You will end with (√5 + 2) / 1 which is simply √5 + 2.

4. Simplify (3 - √2) / (√2 + √3). Multiply by the conjugate (√2 - √3). The numerator becomes (3 - √2) \ (√2 - √3) = 3√2 - 3√3 - 2 + √6. The denominator becomes (√2 + √3) \ (√2 - √3) = 2 - 3 = -1. Putting the simplified numerator over -1 and distributing the negative sign results in -3√2 + 3√3 + 2 - √6.

5. Rationalize 4/(3 - √7). Multiply by the conjugate (3 + √7). The numerator becomes 4(3 + √7) or 12 + 4√7. The denominator becomes (3 - √7) * (3 + √7) = 9 - 7 = 2. The resulting fraction is (12 + 4√7)/2 and simplifies to 6 + 2√7.

6. Simplify √2 / (1 + √2). Multiply by the conjugate (1 - √2). The result will be (√2 - 2) / -1, or 2 - √2.

7. Rationalize 10/(√6 + √4). Multiply by the conjugate (√6 - √4). The numerator is 10(√6 - √4) or 10√6 - 20. The denominator will be (√6 + √4) * (√6 - √4) = 6 - 4 = 2. The rationalized form is (10√6 - 20)/2 which simplifies to 5√6 - 10.

8. Rationalize ∛2/∛3. To simplify, multiply by ∛(3\3)/∛(3\3), resulting in ∛(2*9)/3 or ∛18/3

9. Rationalize 5 / (∛4 + 1). Multiply the fraction by (∛(4\4) - ∛4 + 1)/(∛(4\4) - ∛4 + 1). The numerator is 5(∛16-∛4+1) or 5∛16 - 5∛4 + 5. The denominator is (∛4+1) * (∛16-∛4+1) or 4 - ∛4 +∛4+∛16 -∛4+1

and simplifies to 5 - $\sqrt[3]{4}$+$\sqrt[3]{16}$. Putting them together we have (5$\sqrt[3]{16}$ - 5$\sqrt[3]{4}$ + 5) / (5 - $\sqrt[3]{4}$+$\sqrt[3]{16}$).

These exercises give students a good amount of practice with varied rationalization methods. The questions progress from basic single radical terms to more complicated expressions involving sums and differences and also cube roots. Correctly rationalizing denominators will improve the ability to carry out algebraic manipulations and is an important aspect of algebra and of the EOC exam. In each problem, the objective is to understand when and why to rationalize, and the step-by-step execution of different methods.

Radical Expression Operations Mastery

Performing operations with radical expressions involves a systematic approach to addition, subtraction, multiplication, and division, much like working with polynomial expressions. The key to these operations is understanding how to identify like radicals, combine them, and apply the rules of arithmetic correctly. Like radicals share the same index and radicand, which allows for their coefficients to be added or subtracted. Radicals with different indices or radicands are unlike and cannot be directly combined. When adding or subtracting radical expressions, focus first on identifying terms with identical radicals. For example, consider the expression $3\sqrt{5} + 2\sqrt{5}$. Because both terms include the same radical, $\sqrt{5}$, they can be combined by adding their coefficients: $3 + 2$, yielding $5\sqrt{5}$. However, an expression like $4\sqrt{3} + 2\sqrt{7}$, cannot be simplified further because the radicals are different.

To make things more complicated, sometimes it may appear that radicals are unlike at first, but after simplifying the terms, they can be combined. Take $2\sqrt{12} + 3\sqrt{3}$ for instance. At first glance, it appears they cannot be combined. However, the first term, $\sqrt{12}$, can be simplified by factoring 12 into its perfect square component, 4, and a remaining term, 3, such that $\sqrt{12} = \sqrt{(4\backslash3)} = 2\sqrt{3}$. The expression then transforms into $(2\backslash2\sqrt3) + 3\sqrt3$ which simplifies to $4\sqrt3 + 3\sqrt3$, and then to $7\sqrt3$. Understanding how to decompose radical terms into perfect squares or cubes before attempting to combine them is very important. Also, consider expressions with multiple radicals, such as $\sqrt8 + \sqrt{18} - \sqrt{32}$. Here, each radical can be simplified: $\sqrt8 = \sqrt{(4\backslash2)} = 2\sqrt2$, $\sqrt{18} = \sqrt{(9\backslash2)} = 3\sqrt2$, and $\sqrt{32} = \sqrt{(16\backslash*2)} = 4\sqrt2$. The expression now reads $2\sqrt2 + 3\sqrt2 - 4\sqrt2$, which combines to $(2 + 3 - 4)\sqrt2$, resulting in $1\sqrt2$ or simply $\sqrt2$. Before combining radicals, ensure each radical is reduced to its simplest form. This often involves extracting perfect square or cube factors from the radicand (the value inside the radical symbol).

Multiplication of radical expressions involves similar steps as multiplication of polynomials: the distributive property. For single-term radical expressions, the coefficients and the radicands are multiplied separately. For instance, $2\sqrt3 \backslash 5\sqrt7$ is found by multiplying the coefficients 2 and 5, and the radicands 3 and 7. So, $2 \backslash 5 = 10$ and $3 \backslash 7 = 21$, which yields $10\sqrt{21}$. When multiplying a single term with a multi-term expression, the distributive property applies. For instance, $2\sqrt5(3 + \sqrt2)$ is computed by distributing the $2\sqrt5$ across each term inside the parentheses, yielding $(2\sqrt5 \backslash 3) + (2\sqrt5 \backslash \sqrt2)$. Multiplying this yields $6\sqrt5 + 2\sqrt{10}$. For binomial multiplication of radical expressions, the FOIL method can be used, which stands for First, Outer, Inner, Last. Using the expression $(2 + \sqrt3) \backslash (4 - \sqrt2)$ to illustrate, the first terms $2 \backslash 4 = 8$, the outer terms $2 \backslash -\sqrt2 = -2\sqrt2$, the inner terms $\sqrt3 \backslash 4 = 4\sqrt3$, and the last terms $\sqrt3 \backslash -\sqrt2 = -\sqrt6$. Putting them together gives $8 - 2\sqrt2 + 4\sqrt3 - \sqrt6$. In binomial multiplication, it is very important to check if any of the radical terms may be like radicals to simplify further, as not all the multiplication will lead to unlike radicals. In the case of $(2 + \sqrt3)(2 - \sqrt3)$, the FOIL method yields $4 - 2\sqrt3 + 2\sqrt3 - 3$. The $-2\sqrt3$ and $+2\sqrt3$ cancel, resulting in $4 - 3 = 1$. This example shows how the product of two conjugates of the form $(a + \sqrt b)$ and $(a - \sqrt b)$ results in a rational number, specifically $a^2 - b$. Multiplying radicals also involves recognizing special product forms, which can help make computation faster.

Division of radical expressions often involves rationalizing the denominator, a method detailed in the previous section. Division can be expressed in terms of a fraction such as $(6\sqrt{10}) / (2\sqrt{2})$. The coefficients can be divided separately, so $6/2=3$, and the radicands can be divided separately, so $\sqrt{10} / \sqrt{2} = \sqrt{5}$, resulting in $3\sqrt{5}$. However, if the expression is something such as $1/\sqrt{3}$, the need for rationalization arises, and it involves removing any radicals from the denominator. This is done by multiplying the numerator and denominator by the radical $\sqrt{3}$, resulting in $(1 \setminus \sqrt{3}) / (\sqrt{3} \setminus \sqrt{3}) = \sqrt{3}/3$. As with all algebraic operations, it is important to pay close attention to signs. Mistakes often occur when negative signs are not correctly distributed or combined when adding or subtracting radicals. The process of rationalizing denominators sometimes includes the multiplication of expressions with conjugates, which need careful attention to the rules of distribution.

Another error occurs when incorrectly combining terms that are not like radicals. Terms like $2\sqrt{3}$ and $2\sqrt{5}$ cannot be added or subtracted, and they need to be left in this form. Another common error is prematurely rounding numbers during calculations which may lead to inaccuracies. When solving radical equations, always use exact values until the very last step to keep maximum accuracy. Also, sometimes the expression can be simplified by identifying perfect squares or cubes that were overlooked in the first simplification, so always check if any radical can be further simplified.

Practice problems that involve multiple steps are a good way to assess understanding. For example, consider the expression $2\sqrt{3}(\sqrt{12} - 5\sqrt{2}) + \sqrt{7}(3\sqrt{7} + \sqrt{14})$. The first part can be simplified by distributing, to give $(2\sqrt{3})(\sqrt{12}) - (2\sqrt{3})(5\sqrt{2})$. Since $\sqrt{12} = 2\sqrt{3}$, then this becomes $(2\sqrt{3})(2\sqrt{3}) - (10\sqrt{6})$, or $12 - 10\sqrt{6}$. The second part of the expression becomes $(\sqrt{7})(3\sqrt{7}) + (\sqrt{7})(\sqrt{14})$ which results in $21 + \sqrt{(7\setminus 14)}$ or $21+ \sqrt{(7\setminus 7\setminus 2)}$ or $21 + 7\sqrt{2}$. Putting it all together, this complicated expression simplifies to $12 - 10\sqrt{6} + 21 + 7\sqrt{2}$, or $33 - 10\sqrt{6} + 7\sqrt{2}$. Another good problem is dividing $(4\sqrt{18} - 2\sqrt{24}) / \sqrt{2}$. This expression first needs simplification, specifically $\sqrt{18} = 3\sqrt{2}$, and $\sqrt{24} = 2\sqrt{6}$, so the numerator becomes $4\setminus 3\sqrt{2} - 2\setminus *2\sqrt{6} = 12\sqrt{2} - 4\sqrt{6}$. Dividing this expression by √2, each term gets divided by √2. (12√2)/√2 becomes 12, and (4√6)/√2 = 4√3, so the whole expression becomes 12 - 4√3.

A problem set for students to work through might include the following:

1. Simplify: $3\sqrt{2} + 5\sqrt{2} - 2\sqrt{2}$

2. Simplify: $4\sqrt{5} + 2\sqrt{3} - \sqrt{5} + 6\sqrt{3}$

3. Simplify: $\sqrt{18} + \sqrt{27} - \sqrt{8}$

4. Simplify: $2\sqrt{3}(4\sqrt{6} - \sqrt{2})$

5. Simplify: $(\sqrt{5} + 2)(\sqrt{5} - 3)$

6. Simplify: $(2\sqrt{3} + \sqrt{2})(2\sqrt{3} - \sqrt{2})$

7. Simplify: $(10\sqrt{15}) / (5\sqrt{3})$

8. Simplify: $(6\sqrt{24}) / (3\sqrt{2})$

9. Simplify: $(3 + \sqrt{2}) / \sqrt{2}$

10. Simplify: $4 / (\sqrt{5} - \sqrt{3})$

11. Simplify: $(2\sqrt{3} - 1) / (\sqrt{3} + 1)$

12. Simplify: $\sqrt{8} + \sqrt{12} - \sqrt{18} + \sqrt{27}$

13. Simplify: $2\sqrt{5} (3\sqrt{10} - 2\sqrt{15})$

14. Simplify: $(\sqrt{2} + 3)2 - (\sqrt{3}-2)2$

15. Simplify: $(\sqrt{20} + \sqrt{45})/\sqrt{5}$

16. Simplify: $(2\sqrt{24} - 3\sqrt{54})/\sqrt{6}$

17. Simplify: $\sqrt{2}(\sqrt{8} + 3\sqrt{2}) - (\sqrt{3} + 1)(\sqrt{3} - 2)$

18. Simplify: $(2\sqrt{5} + \sqrt{3})(2\sqrt{5} - \sqrt{3})$

19. Simplify: $4/(\sqrt{3} + \sqrt{2}) - 2/(\sqrt{3} - \sqrt{2})$

20. Simplify: $(2\sqrt{18} - 3\sqrt{8}) / \sqrt{2} + (3\sqrt{27} - 2\sqrt{12})/\sqrt{3}$

These problems are a combination of basic simplifications, operations that include distribution, and expressions involving binomial multiplication. The questions include multiple steps that require combining the concepts of simplification, multiplication, division and rationalization. They will challenge a student to utilize multiple techniques in the same expression. Solving these practice problems will enable a student to better handle complex radical expressions and better prepare for the challenges of the algebra EOC exam. Understanding the concepts behind each calculation will give the student a better chance of success.

Solving Radical Equations Strategically

Radical equations are equations where the variable is located under a radical sign, such as a square root or cube root, presenting unique challenges compared to linear or polynomial equations. Solving these equations requires specific techniques to eliminate the radical and isolate the variable, and a significant hurdle arises from the potential for extraneous solutions, which are solutions obtained through the solving process that do not actually satisfy the original equation. These extraneous solutions occur because squaring both sides of an equation can introduce new solutions, which are not always valid.

A systematic approach to solving radical equations involves three main steps: isolating the radical, squaring both sides of the equation, and checking for extraneous solutions. First, the radical term needs to be isolated on one side of the equation, meaning all other terms should be moved to the opposite side.

This is typically done using inverse operations such as addition, subtraction, multiplication, or division. For example, in the equation $\sqrt{(x+2)} - 3 = 0$, the radical term $\sqrt{(x+2)}$ must be isolated before squaring. To achieve this, add 3 to both sides of the equation resulting in $\sqrt{(x+2)} = 3$.

Next, once the radical is isolated, square both sides of the equation to eliminate the radical sign, which will bring the variable out of the radical. In our example, squaring both sides of $\sqrt{(x+2)} = 3$ gives $(\sqrt{(x+2)})^2 = 3^2$, which simplifies to $x + 2 = 9$. This step transforms the radical equation into a regular polynomial equation, which can now be solved using standard algebraic techniques.

Lastly, it is essential to check all obtained solutions in the original radical equation. This verification step identifies any extraneous solutions that may have been introduced by squaring both sides, because squaring both sides can turn a false equation into a true one. In the case of $x + 2 = 9$, solving for x gives $x = 7$. To verify that this is correct, plug the value of x back into the original equation, $\sqrt{(7+2)} - 3 = 0$. This simplifies to $\sqrt{9} - 3 = 0$, and $3 - 3 = 0$. Since the result is a true statement, $0=0$, $x=7$ is a valid solution. If the equation was something like $\sqrt{(x-3)} = -2$, then squaring both sides would result in $x-3=4$, and $x = 7$, but when x is plugged back into the original equation, $\sqrt{(7-3)} = \sqrt{4} = 2$, which is not equal to -2, which means $x = 7$ is not a valid solution. This highlights the importance of checking for extraneous solutions.

The techniques for solving radical equations are adaptable to different types, including single radical equations, equations with multiple radicals, and more complex scenarios. Single radical equations involve only one radical term in the equation, as in the example above. However, when equations have multiple radical terms, the strategy is a little more intricate. For example, the equation $\sqrt{(x+3)} + \sqrt{(x-2)} = 5$ contains two radical terms, and the isolation of both the radicals is not immediately feasible. In cases like these, the usual strategy is to isolate one of the radicals, square both sides, and repeat the process for the remaining radical terms.

When one side is squared it is critical to remember to use the FOIL method. Take for example $\sqrt{(x + 3)} + \sqrt{(x - 2)} = 5$. Isolate one of the radicals, such as $\sqrt{(x+3)}$ and move the other term to the other side, so $\sqrt{(x+3)} = 5 - \sqrt{(x-2)}$. Now square both sides $(\sqrt{(x+3)})^2 = (5 - \sqrt{(x-2)})^2$. This gives $x + 3 = 25 - 10\sqrt{(x-2)} + x - 2$, which simplifies to $3 = 23 - 10\sqrt{(x-2)}$, moving variables to one side and constants to the other. The equation is $-20 = -10\sqrt{(x-2)}$, and dividing by -10 results in $2 = \sqrt{(x-2)}$. This new equation now contains only one radical, which can be addressed by squaring both sides once more. $4 = x-2$, and therefore $x = 6$. The final step is to check for extraneous solutions: $\sqrt{(6+3)} + \sqrt{(6-2)} = \sqrt{9} + \sqrt{4} = 3 + 2 = 5$, which is the original equation, meaning $x = 6$ is a valid solution.

More complex radical equations require similar strategies but may include additional steps or algebraic manipulations. These types of problems involve more complicated radicands, multiple radical terms, and more complicated expressions. These need a very systematic approach. The main idea is to isolate the radicals one at a time, square both sides, and continue the process, being careful to apply the correct arithmetic and simplification steps. The order of operations (PEMDAS) becomes even more important here. Each step must be checked carefully to avoid errors.

When solving radical equations, there are a few common errors that students tend to make. One common mistake is forgetting to isolate the radical term before squaring both sides. This can lead to more complex and incorrect equations. Also, students may forget to check for extraneous solutions, which will result in invalid answers. Careless application of the distributive property is another source of mistakes, especially when squaring expressions with multiple terms, such as $(a + \sqrt{b})2$, which needs the use of the FOIL method. It is also very easy to make arithmetic errors when simplifying expressions with multiple terms and operations. Students must pay very close attention to each step. Another common error occurs when students fail to simplify radical terms prior to performing operations, which may complicate the process and may even make it impossible to combine terms that would be like radicals if simplified. Also, students may try to perform operations on radicals which have different indices or different radicands which is not allowed.

To illustrate step-by-step solving methods, let's look at another example: $\sqrt{3x + 1} = x - 1$. First, the radical term is already isolated, so next square both sides, $(\sqrt{3x + 1})^2 = (x - 1)^2$. This results in $3x + 1 = x^2 - 2x + 1$. Rearranging terms and setting the equation to zero yields $x^2 - 5x = 0$, and factoring out x, $x(x - 5) = 0$. The solutions are therefore $x = 0$ or $x = 5$. Next, each solution must be plugged back into the original equation to verify validity. With x=0, $\sqrt{3(0) + 1} = 0 - 1$ which means $\sqrt{1} = -1$ which is false. Therefore x=0 is an extraneous solution and not valid. With x=5, $\sqrt{3(5) + 1} = 5 - 1$ becomes $\sqrt{16} = 4$, and 4=4, which is true. Therefore x=5 is the only valid solution.

Let's consider a more complex example with two radicals. Consider $\sqrt{x + 5} - \sqrt{x - 3} = 2$. First, isolate one of the radical terms: $\sqrt{x + 5} = 2 + \sqrt{x - 3}$. Next, square both sides: $(\sqrt{x + 5})^2 = (2 + \sqrt{x - 3})^2$. This results in $x + 5 = 4 + 4\sqrt{x - 3} + x - 3$, which simplifies to $x + 5 = x + 1 + 4\sqrt{x - 3}$. Moving terms around to isolate the radical term, $4 = 4\sqrt{x - 3}$, and dividing by 4 results in $1 = \sqrt{x - 3}$. Squaring both sides gives $1 = x-3$, so $x = 4$. It is now time to check if $x = 4$ is a valid solution. Plugging it into the original equation: $\sqrt{4 + 5} - \sqrt{4 - 3} = \sqrt{9} - \sqrt{1} = 3 - 1 = 2$, which is correct.

The following is a set of problems intended to challenge students' understanding and test their ability to solve radical equations, which will also serve as an overview of the algebra skills developed through this guide:

1. Solve: $\sqrt{2x - 1} = 5$

2. Solve: $\sqrt{3x + 7} = 4$

3. Solve: $\sqrt{x + 3} = x - 3$

4. Solve: $\sqrt{(2x + 5)} = x$

5. Solve: $\sqrt{(4x + 1)} = 2x - 1$

6. Solve: $\sqrt{(x - 2)} + 4 = 7$

7. Solve: $2\sqrt{(x + 3)} - 1 = 9$

8. Solve: $\sqrt{(x + 5)} = \sqrt{(2x - 1)}$

9. Solve: $\sqrt{(3x + 4)} = \sqrt{(x - 2)}$

10. Solve: $\sqrt{(x + 1)} + \sqrt{(x + 6)} = 5$

11. Solve: $\sqrt{(x + 4)} - \sqrt{(x - 1)} = 1$

12. Solve: $\sqrt{(2x + 3)} - \sqrt{(x - 2)} = 2$

13. Solve: $\sqrt{(x + 7)} + \sqrt{(x + 2)} = 5$

14. Solve: $\sqrt{(x + 1)} + x = 5$

15. Solve: $\sqrt{(x + 2)} = x$

16. Solve: $\sqrt{(x - 2)} = x - 4$

17. Solve: $\sqrt{(3x - 2)} = 2\sqrt{(x - 1)}$

18. Solve: $\sqrt{(x + 5)} = 2 - \sqrt{(x)}$

19. Solve: $\sqrt{(x + 3)} + \sqrt{(x - 2)} = 7$

20. Solve: $\sqrt{(2x + 1)} + \sqrt{(x - 3)} = 5$

This set of problems is a mix of single and multiple radical equations. It will test the student's ability to use the principles of isolating radicals, applying correct operations, and identifying extraneous solutions. The problems also require students to integrate skills developed in earlier chapters such as the use of the FOIL method, basic algebraic manipulation, simplifying radical terms and factoring. Solving these problems will give students a very good idea of their capacity and preparedness for the EOC exam and help them gain a better understanding of the algebraic skills they have developed through this guide.

7

DATA ANALYSIS & PROBABILITY

Decoding Data Visualizations

Data visualizations are a key way to understand math and the world around us. Graphs help turn raw numbers into pictures, making it easier to spot patterns, trends, and relationships. Different kinds of graphs are useful for different types of data. Each type presents information in a distinct way. Recognizing the strengths of each graph type is the first step toward becoming proficient in data analysis.

Bar graphs use rectangular bars of varying lengths to show the size of different categories or values. The length of each bar is directly proportional to the value it represents. They're useful for comparing distinct categories or showing how a single value changes over different groupings. For example, a bar graph might show the number of students in each grade at a school. The categories would be the grades, and the height of each bar would represent the number of students in that grade. To get the most out of a bar graph, start by looking at the title, which tells you what data the graph is showing. The labels on the x-axis tell you the categories, and the labels on the y-axis tell you the values. You should also examine the scale of the y-axis to understand the range of values displayed, as a scale that starts at a value other than zero can sometimes make differences seem bigger than they are. Once you understand the axes and scales, you can see the relative size of each category by comparing the lengths of the bars. Longer bars represent larger values, and shorter bars represent smaller values. You can then examine the height of each bar to find the exact value of that category. This approach helps make quick comparisons and find overall trends between the different groups.

Line graphs use points connected by lines to show how a quantity changes over time or across a continuous range. Each point is marked on the graph, and then a line connects the points to create a visual picture of how the data changes. Line graphs are useful for seeing trends, like increases or decreases, and for spotting sudden changes in a variable. For instance, you might use a line graph to chart the temperature over a week, showing how it varies from day to day. To read a line graph effectively, start by identifying the title and the labels on the x-axis (horizontal) and y-axis (vertical). The x-axis often shows time, but it can also represent any continuous variable. The y-axis displays the value of the quantity you're tracking. Look at the scale and intervals on each axis to understand how values are represented. The line connecting the points shows how the value changes from one point to the next. Notice any patterns in the line, such as increases, decreases, or plateaus. Changes in steepness can show how quickly or slowly the values are changing. By analyzing the patterns, slopes, and direction of the line, you can draw conclusions about the data.

Pie charts are circular graphs divided into slices, each representing a proportion of the whole. The size of each slice shows how large that piece is compared to the total. Pie charts are useful when showing how a whole is divided into parts, and they are especially useful for understanding proportions and percentages. For instance, a pie chart can show how a budget is divided among different expenses. Each expense gets its own slice, so it's easy to see which ones take up the most of the budget. To make sense of a pie chart, you have to first look at the title, which will tell you what the chart is about. Each slice will be labeled with a category, and sometimes with a percentage. The size of each slice shows its proportion of the whole. The larger the slice, the larger the proportion it represents. Sometimes, colors are used to make each slice stand out, making the categories easy to tell apart at a quick glance. Using the visual representation of each slice, you can identify which slices are large and which are small, and understand their role in the larger total.

Histograms are similar to bar graphs but are specifically designed to show the distribution of numerical data. Unlike bar graphs, where the categories are distinct and separate, a histogram's bars are connected. They each represent a range of values or 'bins' on a continuous numerical scale. The height of each bar indicates how many data points fall within each range. These graphs are particularly helpful for understanding the frequency of different values. For instance, a histogram can show the distribution of student test scores, indicating how many students scored within different score ranges. When you interpret a histogram, focus on the title and axis labels to understand the type of data and the range of values displayed. The x-axis shows the numerical ranges or bins, and the y-axis indicates the frequency or count of data points in each range. Examine the shape of the distribution to identify any patterns, such as whether the data is centered around a specific value or skewed towards one side. The height of each bar tells you the frequency of data within that range. By studying the shape and distribution, you can find out the common values, any outliers, or any skews.

Successfully reading any kind of graph involves a few key steps. First, you must locate the title, which indicates the graph's main point. It's the first clue to what data the graph is showing. Once you know what the graph is about, focus on the axes. Each one has a label. These labels will help you know what the data being plotted is. For instance, if a graph title is "Sales Over Time", then the x-axis might be labeled "Months" and the y-axis might be labeled "Revenue" in dollars. Next, focus on the scale and intervals used on each axis. Knowing the numerical range of the axes is important for understanding the data. A scale that has been compressed or stretched could show misleading trends. The scale shows what values each mark represents. Once you know the range and intervals of the graph, look for patterns and trends in the data. Trends may be increasing or decreasing values, outliers, and clustering. These patterns may be seen in the shape and direction of lines, the length of bars, or the size of pie slices. Finally, after observing the patterns and trends, use the information to draw conclusions. Use the patterns to show what the data means. Combining all these elements will help you not just read the graphs, but also understand the data they represent.

Now, to put this into practice, consider a bar graph that shows the number of books read by students in different classes. The title is "Books Read Per Class". The x-axis is labeled "Class" and lists different grades like "Grade 9," "Grade 10," "Grade 11," and "Grade 12." The y-axis is labeled "Number of Books," and the scale goes up by intervals of 5, from 0 to 50. You can quickly determine that a higher bar for Grade 10 compared to Grade 12 means that tenth graders read more books on average than twelfth graders. You could then use the scales to get an idea of how many books they each read.

Or, take the example of a line graph. Suppose you see a line graph with the title "Website Traffic Over Time." The x-axis is labeled "Days" and shows the days of the month. The y-axis is labeled "Number of Visitors" and the scale goes from 0 to 1000, with intervals of 100. The line starts at a low point, goes up in the middle, and then levels off at the end. This shows you that website visits rose in the middle of the month and then stopped increasing. You can examine this for more detail using the axes and intervals provided.

Next, consider a pie chart, such as one showing how people spend their time in a day. If the title is "How People Spend a 24-Hour Day," the pie chart could show categories like "Sleep," "Work," "Meals," and "Leisure." Each slice is a part of a whole, and the size shows what part of the whole each slice represents. If the work slice is the biggest, that indicates that work consumes the greatest part of their day. The percentages within each slice are there to show the amount of time they spend on that activity, so that you can easily make comparisons.

Lastly, let's look at a histogram. This histogram shows student scores in a math test. The title is "Math Test Scores." The x-axis has bins, or ranges of scores such as 60-70, 70-80, 80-90, and 90-100. The y-axis shows the number of students falling in each range. If the tallest bar is in the range 80-90, then it means that most students scored in that range. The shapes of the bars can tell you about the skew of the test scores, and help you understand the general distribution.

By combining the ability to read these different graph types, you'll be able to draw conclusions from data that is presented in any form, not just math problems, but also in the world around you. It is important to practice with multiple types of data so that you can confidently read any kind of graph you may encounter. Working through different examples of data visualizations can make you a more skilled data reader, and allow you to more readily understand patterns that might not otherwise be apparent from tables of numbers.

Central Tendency Mastery

Central tendency measures are key tools for summarizing and understanding data sets, offering a way to pinpoint a typical or central value within a group of numbers. Three commonly used measures of central tendency are the mean, the median, and the mode. Each measure offers a unique perspective on the center of a dataset and is appropriate for different circumstances. The range, while not a measure of central tendency, provides context to the spread of data around that center.

The mean, often called the average, is calculated by adding up all the values in a dataset and dividing by the total number of values. For example, if a dataset consists of the numbers 2, 4, 6, 8, and 10, the mean would be calculated as $(2 + 4 + 6 + 8 + 10) / 5$, which equals $30 / 5$, or 6. The mean represents the balancing point of the data, where the sum of the distances from each point to the mean is zero. A major strength of the mean is that it uses every value in the data, making it sensitive to changes in any value. However, this sensitivity also means that the mean is susceptible to outliers. Outliers are extremely high or low values that can pull the mean away from the center of the bulk of the data. For instance, if you were to add the value 50 to the dataset above, the new mean would be (2 + 4 + 6 + 8 + 10 + 50) / 6, which equals 80 / 6, or about 13.33. The presence of a single outlier has shifted the mean significantly higher. This characteristic of the mean is important to understand when deciding if it is an appropriate representation of the central tendency.

The median, unlike the mean, focuses on the position of values within a dataset. To find the median, the data must first be ordered from least to greatest. If there is an odd number of data points, the median is the middle value. If there is an even number of data points, the median is the mean of the two middle values. Using the earlier dataset of 2, 4, 6, 8, and 10, the median is 6, as it's the middle value when the dataset is ordered. For the dataset with the outlier, 2, 4, 6, 8, 10, and 50, the median would be (6+8)/2, or 7. Compared to the drastic shift in the mean when adding the outlier, the median barely shifted. The median is resistant to outliers and extreme values, because it's determined by the order of values rather than the actual values themselves. This makes it a useful central tendency measure when a data set is suspected to have extreme values. It's also more suited to skewed distributions, where most values are clustered on one end with a few extreme values on the other.

The mode identifies the value or values that appear most frequently in a data set. Unlike the mean and median, which require numerical data, the mode can also be used with categorical data. In a data set like 2, 4, 4, 6, 8, the mode is 4 because it appears twice. There may be more than one mode if multiple values appear with the same greatest frequency. A dataset with two modes is called bimodal, and one with multiple modes is called multimodal. If every value appears only once, the data set is considered to have no mode. The mode is helpful in understanding the most common categories, values or events in a data set. While not as robust as the mean and median for numerical data, it is a useful tool for categorical data or for identifying typical values in a dataset.

The range is not a measure of central tendency; however it provides important context by giving an indication of how spread out the data is. It is calculated as the difference between the maximum and minimum values in the dataset. Using the set 2, 4, 6, 8, and 10, the range would be 10 - 2, or 8. For the data set with the outlier, 2, 4, 6, 8, 10, and 50, the range would be 50 - 2, or 48. While the range doesn't tell us about the center of the data, it is helpful to see how large or small the spread of the data is.

Choosing the best measure of central tendency depends on the type of data and the goal of the analysis. The mean is typically used with numerical data that is evenly distributed. This may include scores on a standardized test, the average age of students in a school or temperatures during a week. It is also sensitive to all changes in the data, which may be a good or a bad thing, depending on the specific needs of the analysis. When a data set is not evenly distributed, or has suspected outliers, the median is often the more appropriate measure. Real-world examples of when to use the median would include income distribution, the price of houses in a city, or the sizes of different organizations. These measures would be skewed, and the median provides a more realistic typical value. The mode is helpful when you want to know what the most common category is, such as a person's favorite color or common brands of cars.

Outliers can greatly affect the central tendency measures, and the impact of outliers can have real effects in the world. Consider a dataset of the average income in a neighborhood. Suppose there is one very wealthy person with a much higher income than everyone else. This one person could significantly pull the average income up, making it seem like people are better off than they are. In this situation the median income would provide a much more realistic perspective of the typical income of that neighborhood, due to its resistance to outliers. Consider, instead, a different scenario of a teacher giving a test. Suppose a few students got extremely high scores, while the others got typical scores. In this case, the average score would be a reasonable estimate of the class's central performance. In general, it is wise to look at both the mean and median to fully understand the typical value in a dataset. When these measures are the same or close, then it means the data is likely evenly distributed, without outliers. But when they are different, this indicates that either outliers are present, or that the data is skewed.

Consider a real-world example in education. A teacher wants to understand how well her students performed on a recent exam. She could use the mean to calculate the average score, providing a measure of overall class performance. However, if there are a few students who did exceptionally poorly or exceptionally well, these outlier scores could distort the mean, potentially not reflecting the performance of the majority of the class. In this case, the median would give a better idea of where most of the students are scoring because it's less affected by outliers. The teacher may also consider using the mode if she wants to see what the most common score is in the class. The range of scores would help the teacher to know how large the spread of scores was, and how much variability there was from the lowest to highest score.

In business, the mean, median, mode, and range all have their uses. When a company wants to calculate its average monthly revenue, it might use the mean as a basic measure. But if the company has a few months with very high or very low revenue, it might also examine the median revenue to get a more stable idea of its typical revenue without the effect of those outliers. The mode might be used to see the most common product sold, or the most common size of product that is bought. The range of revenue helps the business understand how volatile its revenue is, and whether it can make accurate predictions for future financial planning.

In healthcare, these measures are used to understand patient data. For example, doctors might examine the mean blood pressure for patients in a certain age group, but if there are extreme values, the median would give them more information. They could examine the mode to see the most common diagnosis seen at a certain clinic, or the range of patient ages to understand the age groups they serve.

In summary, these statistical measures provide different lenses through which to view the central tendency of a dataset. The mean uses every value in a dataset to measure the average; however it is sensitive to outliers. The median finds the middle point of ordered data, is resistant to outliers, and is more suitable for skewed data sets. The mode identifies the most frequent value, and is useful for categorical data. And the range gives context to how spread out the values are in a data set. Each measure has unique advantages and uses, and selecting the proper measure requires an understanding of the data and the goals of the analysis. By combining an understanding of these measures, it's possible to get a better and more complete picture of the patterns in any dataset.

Probability Fundamentals Explored

Probability is the measure of how likely an event is to occur, expressed as a number between 0 and 1, where 0 means the event is impossible, and 1 means the event is certain. To start, understanding probability involves grasping a few basic terms. An event is a specific outcome or set of outcomes of a situation, like flipping a coin and getting heads or rolling a die and getting a 4. A sample space is the set of all possible outcomes. For example, when flipping a coin, the sample space is {heads, tails}, and when rolling a standard six-sided die, the sample space is {1, 2, 3, 4, 5, 6}. The probability of an event is calculated by dividing the number of favorable outcomes by the total number of possible outcomes. For example, the probability of rolling a 4 on a six-sided die is 1/6 because there is only one way to get a 4, and six possible outcomes. The probability of an event is often written as P(event). Therefore, P(rolling a 4) = 1/6.

Probabilities can be written as fractions, decimals, or percentages. A probability of 1/2 is the same as 0.5 or 50%. The closer a probability is to 1, the more likely the event is to happen. The probability scale always goes from 0 to 1. Understanding this is fundamental for making predictions. In contrast, a probability near 0 indicates the event is not likely to happen. For example, in a standard deck of 52 cards, the probability of drawing an ace is 4/52, or 1/13, or about 7.7%, which is a lower probability than, for example, drawing any heart at 13/52, or 1/4 or 25%. Calculating these probabilities helps to measure and make predictions about future events.

Simple probability deals with the likelihood of a single event. Compound probability, on the other hand, considers multiple events and their combinations. For a compound event, the 'and' rule, also known as the multiplication rule, applies when one wants to find the probability of two or more independent events occurring. Independent events are events where the outcome of one does not affect the outcome of the others, such as flipping a coin multiple times. The probability of two independent events happening is found by multiplying their individual probabilities. For instance, the probability of flipping a fair coin twice and getting heads both times is (1/2) * (1/2) = 1/4. So, the probability of getting two heads in a row is 1/4. This is often used in games and lotteries, where the events of each draw or roll are independent of each other.

The 'or' rule, also known as the addition rule, helps when you need to determine the probability of one or another event occurring. For mutually exclusive events, events that can not occur at the same time, the probability is the sum of their individual probabilities. For example, if you were to roll a die, the probability of rolling a 1 or 2 is found by adding the probability of each individual event. Because these events are mutually exclusive, P(rolling a 1 or a 2) = P(rolling a 1) +

P(rolling a 2) = 1/6 + 1/6 = 2/6, or 1/3. However, if the events are not mutually exclusive, you have to subtract the probability of both occurring to avoid over counting, because the probability of both occurring has been added in twice. For example, the probability of drawing either a heart or a face card from a deck of cards. These are not mutually exclusive, because it is possible to draw a face card that is also a heart. So, P(heart or face card) = P(heart) + P(face card) - P(heart and face card). Because the probability of drawing a heart is 13/52, the probability of drawing a face card is 12/52, and the probability of drawing a face card that is also a heart is 3/52, P(heart or face card) = 13/52 + 12/52 - 3/52 = 22/52 or 11/26.

Sample spaces provide an overview of all possible outcomes in a given situation. These can be represented as lists, tables, or diagrams, and they help ensure that all outcomes are considered in calculating probability. For example, when considering the sample space of two coin flips, you can write the outcomes as a list: {HH, HT, TH, TT} where H represents heads and T represents tails. Or you could show it as a table:

	Heads	Tails
Heads	HH	HT
Tails	TH	TT

This visual representation makes it easy to see there are four equally likely outcomes. Understanding sample spaces allows for a systematic method to find probabilities.

Theoretical probability relies on understanding the situation and determining the probability based on reasoning, rather than conducting experiments. It presumes a perfect world and ideal outcomes, often based on mathematical reasoning. For instance, when rolling a fair six-sided die, the theoretical probability of getting a 3 is 1/6 because each side of the die has an equal chance of landing up. Or in a deck of cards, the theoretical probability of drawing a spade is 13/52, because there are 13 spades in a deck of 52 cards, assuming they all have an equal chance of being drawn. These are probabilities that do not rely on conducting experiments, but are derived by logical reasoning based on the properties of the event.

Probability rules help to perform more complex calculations, like the multiplication rule for independent events, and the addition rule for mutually exclusive events. These rules are essential when dealing with multiple events. For example, a bag contains 4 red marbles and 6 blue marbles. If you draw one marble, then return it, then draw again, what's the probability of drawing a red marble both times? Since the marbles are returned, the events are independent. The probability of drawing a red marble the first time is 4/10. The probability of drawing a red marble a second time is also 4/10. Using the multiplication rule, the probability of drawing a red marble twice is (4/10) * (4/10), or 16/100. Applying these rules allows for the prediction of outcomes in compound probability scenarios.

Conditional probability deals with the probability of an event given that another event has already occurred. This introduces a twist in the calculations because it limits the sample space. This is different than independent events, where the outcome of one event does not impact the next. Conditional probability, however, is written as P(A|B), which means the probability of event A given that event B has already occurred. A common example is drawing cards from a deck without replacement. If you draw a card from a standard deck of cards and do not replace it, the probability of drawing another card changes. For example, the probability of drawing a king from a deck of cards is 4/52. But if we draw a card and do not replace it, and that card was a king, the probability of the next card being a king is now 3/51, because there are only 3 kings left, and a total of 51 cards remaining. So the probability of drawing a second king is conditional on having drawn a first king. Conditional probability helps us deal with scenarios where prior outcomes affect subsequent events.

Solving multi-step probability problems involves combining all the previous concepts. These problems might involve a series of independent events, conditional probabilities, or both. For example, consider a situation where you flip a coin and then draw a card. What is the probability of flipping heads, then drawing a face card? You need to apply both the multiplication rule for independent events and also calculate both the probability of flipping heads, which is 1/2, and the probability of drawing a face card, which is 12/52. Using the multiplication rule, P(heads and face card) is equal to (1/2)*(12/52), or 12/104, which simplifies to 3/26. Multi-step problems require a systematic approach, and understanding the underlying probabilities.

Probability is not just for classroom problems; it has applications in numerous fields. In games and gambling, understanding probability is essential for calculating odds and making decisions. Players weigh the likelihood of different outcomes, which helps to inform their bets. In sports, probability is used to predict game outcomes, analyze player performance, and make strategic decisions. For example, coaches analyze the probability of a successful free throw or a successful pass to make informed choices during games. Probability also plays a large part in weather forecasting, where meteorologists use probability to predict the likelihood of rain or snow, allowing people to plan for the coming weather conditions. The probability of a specific weather condition helps people to prepare for those outcomes. And in insurance, companies use probability to assess risk and determine premiums. They use historical data to calculate the probability of accidents or other events to set appropriate costs for their insurance policies.

Probabilistic thinking is a way to approach decision making by weighing the chances and evaluating options based on likelihood rather than certainty. This involves analyzing possible outcomes, determining their probabilities, and making a choice based on the likelihood of achieving a desired result. For example, consider a person debating whether or not to take a new job. They would look at the probability of job security in the new job, as well as the probability of the new job offering higher pay, or more opportunities, and use these probabilities to help make an informed decision. It does not guarantee a positive result, but it enables people to weigh options based on the most probable outcome. Probabilistic thinking enables more informed, data-driven decisions in everyday life, rather than decisions based on intuition alone.

Probability is a powerful tool for analyzing uncertainty and making informed decisions. By understanding basic concepts, mastering calculations, and seeing practical applications, abstract mathematical notions turn into tangible, applicable skills. These basic principles of probability become useful in games, sports, and daily life.

Correlation and Data Relationships

Data analysis often involves examining how different variables relate to one another, which is where scatter plots and correlation analysis become useful tools. A scatter plot is a graph that visually displays the relationship between two numerical variables. Each point on the graph represents a pair of values for these variables. The x-axis represents one variable, and the y-axis represents the other. By looking at the pattern of points on the scatter plot, you can gain insight into how these two variables might be related. If there is a pattern, this suggests a relationship may exist between the two variables, but if the points seem randomly scattered, this suggests there is no linear relationship.

Scatter plots help to identify the direction and strength of a relationship. In a positive correlation, as one variable increases, the other variable also tends to increase. On a scatter plot, this would be shown as points generally moving upward and to the right. For example, a scatter plot might show a positive correlation between the number of hours a student studies and their test scores. As study hours increase, test scores tend to increase. A negative correlation, by contrast, is when one variable increases, the other variable tends to decrease. This would be shown on the scatter plot as points generally moving downward and to the right. For example, there might be a negative correlation between the price of a product and the quantity demanded. As prices go up, quantity demanded tends to decrease. If a relationship exists, but there is no clear upward or downward trend, the variables have no correlation. This would be shown by a random scattering of points across the plot, rather than a pattern. For example, there would likely be no correlation between a person's shoe size and their score on an algebra test.

The strength of a correlation refers to how closely the points on the scatter plot follow a trend. A strong correlation has points that cluster closely around a line, either a positively sloped line or a negatively sloped line. The more tightly clustered the points, the stronger the correlation. A weak correlation shows points that are more scattered away from any potential line. The more dispersed the points, the weaker the correlation. If there is no line pattern at all, the correlation is zero. The strength of a correlation does not indicate the steepness of the relationship, but how closely the points fit the pattern. So a strong negative correlation could be a less steep slope than a weak positive correlation; strength is about how closely the points follow a linear trend.

Correlation also has a direction, which can be positive or negative. Positive correlations, where both variables move in the same direction, occur in many real world scenarios. For example, as temperature increases, ice cream sales tend to increase as well. Negative correlations show that as one variable goes up, the other goes down. For example, as the amount of sleep someone gets decreases, their alertness during the day will probably also decrease. These relationships can be seen on a scatter plot and analyzed for strength as well.

A key concept to understand is the difference between correlation and causation. Correlation means that two variables move together, either in the same direction (positive correlation) or in opposite directions (negative correlation). Causation, however, means that changes in one variable directly cause changes in the other variable. While correlation can point to a possible relationship between variables, it doesn't mean that one variable is the cause of changes in the other. It is possible for two variables to be correlated, but not causally linked. For example, there may be a correlation between ice cream sales and the number of crimes reported. As ice cream sales go up, so does the crime rate. However, eating ice cream does not cause crime. Both of these things probably go up in the summer because the weather is warmer. This is known as a confounding variable, or a third variable, which affects both the variables being measured. Confounding variables can make it look like there is a causal relationship when there is not one.

To further clarify, consider this example: a study might reveal a positive correlation between the number of firefighters sent to a fire and the amount of damage caused by the fire. Does this mean that the firefighters caused the damage? Obviously not. The explanation is that larger fires require more firefighters and cause more damage, making the size of the fire a confounding variable that explains the correlation between number of firefighters and the amount of damage. So the number of firefighters does not cause the increase in damage. Because of these types of situations, correlation alone is not enough to prove causation. There must be other types of evidence to support a claim of causation. This is why carefully designed experiments, where confounding variables are controlled or eliminated, are necessary to prove causation.

In addition to simply looking at a scatter plot, we can also calculate a correlation coefficient, which is a numerical value that measures the strength and direction of a linear relationship between two variables. The most common correlation coefficient is the Pearson correlation coefficient, also known as Pearson's r. This value will always be a number between -1 and +1. A value of +1 represents a perfect positive correlation, and all the points on the scatter plot will perfectly fit a straight, upward sloping line. A value of -1 represents a perfect negative correlation, and all the points will perfectly fit a downward sloping line. A value of 0 indicates no linear correlation at all, and the points will seem randomly scattered. Values between 0 and 1 represent positive correlations of varying strengths, and values between 0 and -1 represent negative correlations of varying strengths.

For the Pearson correlation coefficient, the values closer to +1 or -1 represent stronger correlations, and those closer to 0 represent weaker correlations. A correlation coefficient of 0.8, for example, is considered a strong positive correlation, while a coefficient of -0.2 would indicate a weak negative correlation. There are rules of thumb for determining the strength of the coefficient. The correlation coefficient of ±0.7 or higher is considered a strong correlation, between ±0.5 and ±0.7 is considered a moderate correlation, and from 0 to ±0.5 is considered a weak correlation. This measure helps to quantify the linear relationship you see on a scatter plot, to make the relationship more clear, objective, and comparable across datasets.

To calculate the Pearson correlation coefficient, one must follow a series of steps using the values of each paired data point of x and y variables. First, you calculate the mean (average) of all the x values, which is referred to as $\bar{x}$, and you also calculate the mean of all the y values, which is referred to as $\bar{y}$. Then, you determine how far each x value is from the mean x value, so the difference between each x and $\bar{x}$, which is referred to as $x-\bar{x}$. The same is done for all y values, calculating the difference between each y and $\bar{y}$, which is referred to as $y-\bar{y}$. You then multiply the values $x-\bar{x}$ and $y-\bar{y}$ for each pair of data points, and then you sum all these multiplied values together. You must also calculate the squared differences of each of the $x-\bar{x}$ values, and then sum them. And you must do the same with the $y-\bar{y}$ values, calculating the squared differences, and then summing them. Finally, you divide the summed multiplied values by the square root of the summed squared x difference multiplied by the summed squared y difference. So, after following all of these complex calculations, you have Pearson's r. Because it is rather complex, the Pearson correlation coefficient is often calculated with calculators or computer programs. However, the purpose of the complex calculation is to arrive at a single number that provides a clear, objective analysis of the linear relationship between two numerical variables.

Creating a scatter plot is a relatively simple process, especially when using graphing software or spreadsheets, but knowing when to use them and how to interpret them is important. You first need to have your data, which consists of paired values of two variables. You then choose which variable goes on the x axis, and which goes on the y axis. Typically, the variable that you believe is causing the changes in the other variable goes on the x axis, because that's what is being manipulated in controlled experiments. The dependent variable goes on the y axis because it depends on the value of the independent variable. You then plot each data point by making a dot on the graph corresponding to the x and y values. Once all the data points are plotted, you can visually examine the scatter plot to see if any patterns or trends appear. If the points show a linear pattern, then a correlation may exist between the variables. The direction of that correlation will be seen through the slope of any linear relationship, and the strength will be indicated by how tightly grouped the points are. If the points seem scattered with no clear pattern, then there is no linear relationship. Understanding that correlation is not causation and being aware of possible confounding variables will allow you to analyze the data more completely.

Understanding scatter plots and correlation analysis provides useful tools for exploring data relationships. By creating scatter plots, identifying correlation types, interpreting the strength and direction of correlations, and avoiding the mistake of assuming causation, you can effectively analyze data and draw well-informed conclusions. It provides the basis for further statistical analysis, allowing for a deeper understanding of complex data sets and how variables interact in real world situations. These skills allow a more nuanced and clear picture of statistical data.

8.1 FULL-LENGTH PRACTICE TEST 1

Section 1: Algebra Foundations

Properties of Real Numbers

1. Which of the following is an example of the **distributive property**?

A) $a+b = b + a$

B) $a(b+c) = ab + ac$

C) $a(bc) = (ab)c$

D) $a+(b+c) = (a+b)+c$

Order of Operations (PEMDAS)

2. Solve the following expression:

$8+4\times(6-2)\div2$

A) 14

B) 16

C) 18

D) 20

Evaluating Algebraic Expressions

3. If $x=3$ and $y=2$, evaluate:

$5x^2-3y+4$

A) 38

B) 40

C) 43

D) 44

Simplifying Expressions

4. Simplify:

$(3x-2)+(5x+6)$

A) $8x+4$

B) $8x-4$

C) $15x+4$

D) $15x-4$

Solving Basic Equations & Inequalities

5. Solve for x:

$5x-3=2x+9$

A) 2

B) 4

C) 6

D) 8

Understanding Absolute Values

6. Solve:

$|3x-7|=5$

A) $x=4, x=2/3$

B) $x=2, x=4/3$

C) x=3,x=2

D) x=4,x=1

Section 2: Linear Equations & Functions

Understanding Functions (Definition & Notation)

7. Which of the following represents a function?

A) $y=x^2$

B) $x^2+y^2=4$

C) $x=y^2$

D) $x^2+y=7$

Graphing Linear Equations (Slope & Intercepts)

8. What is the **slope** of the line passing through points (3,2) and (7,6)?

A) 1

B) 2

C) 3/2

D) 4/3

Writing Linear Equations (Slope-Intercept, Point-Slope, & Standard Form)

9. Find the equation of the line in **slope-intercept form** that passes through (2,5) with a slope of 3.

A) $y=3x+1$

B) $y=3x-1$

C) $y=3x+5$

D) $y=3x-5$

Section 2: Linear Equations & Functions (Continued)

Solving Systems of Linear Equations (Graphing, Substitution, & Elimination)

10. Solve the system of equations using substitution:

y=2x+1, 3x−y=5.

A) (2,5)
B) (3,7)
C) (1,3)
D) (4,9)

Modeling Real-World Problems with Linear Equations

11. A store charges a $5 entrance fee and $2 per item purchased. If **x** represents the number of items bought, which equation represents the total cost C?
A) C=5x+2
B) C=2x+5
C) C=5(x+2)
D) C=2(x+5)

Section 3: Quadratic Equations & Functions

Understanding Quadratic Functions & Their Graphs

12. Which of the following represents the graph of a quadratic function?
A) A straight line
B) A V-shaped graph
C) A U-shaped curve (parabola)
D) A zig-zag pattern

Solving Quadratic Equations (Factoring, Quadratic Formula, Completing the Square)

13. Solve the equation:

$$x^2 - 5x + 6 = 0$$

A) $x = 3, x = 2$
B) $x = -3, x = -2$
C) $x = 6, x = -1$
D) $x = 5, x = -1$

Applications of Quadratics in Real Life

14. A ball is thrown upward with a height function given by:

$$h(t) = -16t^2 + 48t + 5$$

What is the **initial height** of the ball?
A) 0 feet
B) 5 feet
C) 16 feet
D) 48 feet

Quadratic Inequalities

15. Solve the inequality:

$$x^2 - 4x - 5 > 0.$$

A) $x < -1$ or $x > 5$
B) -1
C) $x < -5$ or $x > 1$
D) -5

Section 4: Exponential Functions & Properties

Exponential Growth & Decay

16. The population of a city is modeled by:

$P(t)=5000(1.05)^t$

What does the **1.05** represent?
A) Initial population
B) Growth rate
C) Time in years
D) Final population

Section 5: Polynomials & Factoring

Adding, Subtracting, Multiplying, & Dividing Polynomials

17. Simplify:

$(3x^2+2x-1)-(x^2-4x+5)$

A) $2x^2+6x-6$
B) $4x^2-2x+4$
C) $2x^2+6x+4$
D) $2x^2-2x-6$

Section 6: Radical & Rational Expressions

Simplifying Square Roots & Cube Roots

18. Simplify:

$\sqrt{50}$

A) $5\sqrt{2}$
B) $10\sqrt{5}$
C) $25\sqrt{2}$
D) $2\sqrt{5}$

Section 7: Data Analysis & Probability

Mean, Median, Mode & Range

19. Find the **mean** of the numbers: 3,7,7,9,10.

A) 6

B) 7.2

C) 8

D) 9

Section 7: Data Analysis & Probability (Continued)

Probability Rules & Applications

20. A box contains 4 red, 3 blue, and 5 green marbles. If a marble is drawn at random, what is the probability of selecting a **blue** marble?

A) 3/12

B) 1/12

C) 1/3

D) 3/10

Scatter Plots & Correlation

21. If a scatter plot shows that **as x increases, y decreases**, what type of correlation does the data have?

A) Positive correlation

B) Negative correlation

C) No correlation

D) Undefined correlation

Section 6: Radical & Rational Expressions (Continued)

Rationalizing Denominators

22. Simplify:

$5/\sqrt{3}$

A) $5\sqrt{3}/3$
B) $5/3\sqrt{3}$
C) $5\sqrt{3}/6$
D) $5\sqrt{3}/5$

Adding, Subtracting, Multiplying, & Dividing Radical Expressions

23. Simplify:

$(2\sqrt{5}+3\sqrt{2})-(\sqrt{5}+2\sqrt{2})$.

A) $\sqrt{5}+\sqrt{2}$
B) $\sqrt{5}-\sqrt{2}$
C) $\sqrt{10}+\sqrt{4}$
D) $2\sqrt{5}+\sqrt{2}$

Solving Radical Equations

24. Solve for x:

$\sqrt{x}+3=5$.

A) $x=22$
B) $x=25$
C) $x=28$
D) $x=30$

Section 5: Polynomials & Factoring (Continued)

Factoring Techniques (GCF, Difference of Squares, Trinomials)

25. Factor completely:

x^2-16

A) $(x-4)(x+4)$
B) $(x-2)(x+8)$
C) $(x-8)(x+2)$
D) $(x-4)(x-4)$

Solving Polynomial Equations

26. Solve:

$x^3-8=0$

A) $x=2$
B) $x=-2$
C) $x=4$
D) $x=-4$

Section 4: Exponential Functions & Properties (Continued)

Simplifying & Solving Exponential Equations

27. Solve for x:

$2^{x+1}=16.$

A) $x=3$
B) $x=4$
C) $x=5$
D) $x=6$

Comparing Linear & Exponential Functions

28. Which function **grows faster** over time?

A) f(x)=2x+3

B) g(x)=3^x

C) h(x)=x^2−1

D) j(x)=x/2

Final Probability & Data Questions

Interpreting Graphs & Tables

29. What type of graph is best for **showing trends over time**?

A) Bar graph

B) Line graph

C) Pie chart

D) Histogram

Mean, Median, Mode, & Range (Advanced)

30. Find the **median** of the numbers: 5,8,12,15,22,25,30.

A) 12

B) 15

C) 18

D) 20

Section 1: Algebra Foundations

Properties of Real Numbers

31. Which of the following correctly demonstrates the **associative property of multiplication**?

A) (2×3)×4=2×(3×4)

B) 5+0=5

C) 6×1=6

D) 4+3=3+4

Order of Operations (PEMDAS)

32. Evaluate the expression:

$10+4×3−6÷2$

A) 14

B) 16

C) 189

D) 20

Evaluating Algebraic Expressions

33. If a=−2 and b=4, evaluate:

$2a^2−3b+5$

A) -5

B) 1

C) 3

D) 0

Simplifying Expressions

34. Simplify:

$(4x+5)−(2x−3).$

A) 2x+8

B) 6x−2

C) 2x+2

D) 6x+8

Solving Basic Equations & Inequalities

35. Solve for x:

$7x - 4 = 3x + 8$.

A) $x = 2$
B) $x = 3$
C) $x = 4$
D) $x = 5$

Understanding Absolute Values

36. Solve:

$|2x - 1| = 5$

A) $x = 3, -2$
B) $x = 2, -1$
C) $x = 4, -2$
D) $x = 6, -3$

Section 2: Linear Equations & Functions

Understanding Functions (Definition & Notation)

37. Which relation is **not** a function?
A) $y = x^2 + 3$
B) $x = y^2 - 1$
C) $y = 2x + 4$
D) $y = 5/x$

Graphing Linear Equations (Slope & Intercepts)

38. Find the **y-intercept** of the line given by:

$3x-2y=6$

A) $(0,-3)$
B) $(0,3)$
C) $(0,-2)$
D) $(0,2)$

Writing Linear Equations (Slope-Intercept, Point-Slope, & Standard Form)

39. What is the **slope-intercept form** of a line passing through $(4,7)$ with slope $m=2$?
A) $y=2x+3$
B) $y=2x-1$
C) $y=2x+1$
D) $y=2x-4$

Section 3: Quadratic Equations & Functions

Understanding Quadratic Functions & Their Graphs

40. Which of the following represents the standard form of a quadratic equation?
A) $y=ax+b$
B) $y=ax^2+bx+c$
C) $y=a(x-h)^2+k$
D) $y=mx+b$

Solving Quadratic Equations (Factoring, Quadratic Formula, Completing the Square)

41. Solve the quadratic equation:

$x2-7x+12=0$

A) x=3,x=4

B) x=−3,x=−4

C) x=6,x=2

D) x=7,x=5

Applications of Quadratics in Real Life

42. A ball is thrown into the air, and its height is modeled by:

$h(t)=-5t^2+20t+3.$

What is the maximum height the ball reaches?
A) 18 meters
B) 20 meters
C) 23 meters
D) 25 meters

Quadratic Inequalities

43. Solve the inequality:

$x^2-5x+6<0$

A) −2
B) 2
C) x<−2 or x>3
D) x<2 or x>3

Section 4: Exponential Functions & Properties

Exponential Growth & Decay

44. A population of bacteria grows according to the equation:

$P(t)=200(1.08)^t$

What does the **1.08** represent?

A) Growth rate of 8%

B) Initial population

C) Doubling time

D) Decay factor

Laws of Exponents

45. Simplify:

$(3x^2y^3)^2$

A) $6x^4y^6$
B) $9x^4y^6$
C) $3x^4y^6$
D) $9x^2y^3$

Simplifying & Solving Exponential Equations

46. Solve for x:

$5^{x+1}=125$

A) x=1
B) x=2
C) x=3
D) x=4

Comparing Linear & Exponential Functions

47. Which function grows **faster** over time?

A) f(x)=4x+5

B) g(x)=2^x

C) $h(x)=x^2+3$

D) $j(x)=x^3$

Real-World Exponential Function Application

48. The value of a car depreciates according to the function:

$V(t)=25,000(0.85)^t$

What does **0.85** represent?

A) A 15% increase each year

B) A 15% depreciation rate

C) The car's final value

D) The car's original price

Section 5: Polynomials & Factoring

Adding, Subtracting, Multiplying, & Dividing Polynomials

49. Simplify:

$(2x^2+3x-4)+(5x^2-2x+6).$

A) $7x^2+x+10$

B) $3x^2+5x+2$

C) $7x^2+x+2$

D) $7x^2+5x+10$

Factoring Techniques (GCF, Difference of Squares, Trinomials)

50. Factor completely:

$6x^2-11x+4.$

A) $(2x-1)(3x-4)$

B) $(3x-1)(2x-4)$

C) $(2x-3)(3x-4)$

D) $(2x-2)(3x-3)$

Solving Polynomial Equations

51. Solve for x:

$x^3-27=0$

A) $x=3$

B) $x=-3$

C) $x=9$

D) $x=27$

Section 6: Radical & Rational Expressions

Simplifying Square Roots & Cube Roots

52. Simplify:

$\sqrt{75}$

A) $5\sqrt{3}$

B) $3\sqrt{5}$

C) $25\sqrt{3}$

D) $2\sqrt{15}$

Rationalizing Denominators

53. Simplify:

$7/\sqrt{2}$

A) $7\sqrt{2}/2$

B) $14/\sqrt{2}$

C) $7/2\sqrt{2}$

D) $\sqrt{2}/7$

Adding, Subtracting, Multiplying, & Dividing Radical Expressions

54. Simplify:

$(4\sqrt{3}+2\sqrt{5})-(2\sqrt{3}+\sqrt{5})$.

A) $2\sqrt{3}+\sqrt{5}$

B) $2\sqrt{3}+3\sqrt{5}$

C) $6\sqrt{3}+\sqrt{5}$

D) $2\sqrt{3}-\sqrt{5}$

Solving Radical Equations

55. Solve for x:

$\sqrt{x}+7=5$.

A) x=18

B) x=25

C) x=20

D) x=15

Section 7: Data Analysis & Probability

Interpreting Graphs & Tables

56. A company tracks its monthly sales and displays the data in a bar graph. Which feature of the graph helps determine the highest sales month?
A) The color of the bars

B) The height of the bars

C) The labels on the x-axis

D) The number of bars

Mean, Median, Mode & Range

57. Find the **mean** of the following set of numbers:

4,8,10,12,16

A) 8

B) 10

C) 12

D) 14

Identifying the Median in a Data Set

58. What is the median of the following numbers?

5,11,7,9,15

A) 7

B) 9

C) 11

D) 15

Identifying the Mode in a Data Set

59. What is the mode of the numbers:

3,7,8,3,9,8,8,5

A) 3

B) 5

C) 7

D) 8

Finding the Range of a Data Set

60. Find the **range** of the following set of numbers:

6,12,4,15,9

A) 9

B) 11

C) 13

D) 15

Probability Rules & Applications

61. A bag contains 5 red, 3 blue, and 7 green marbles. What is the probability of drawing a **blue marble**?

A) 3/12

B) 3/15

C) 5/15

D) 715

Theoretical Probability vs. Experimental Probability

62. A coin is flipped 50 times, and heads appears 28 times. What is the **experimental probability** of flipping heads?

A) 22/50

B) 25/50

C) 28/50

D) 30/50

Probability of Independent Events

63. A die is rolled, and a coin is flipped. What is the probability of rolling a **3** and flipping **tails**?

A) 1/12

B) 1/6

C) 1/2

D) 1/3

Probability of Dependent Events

64. A box contains 4 red, 3 blue, and 5 green balls. If one ball is drawn and **not replaced**, then another is drawn, what is the probability that both are red?

A) 2/15

B) 3/15

C) 1/5

D) 2/5

Scatter Plots & Correlation

65. A scatter plot shows that as **the number of hours studied increases, test scores also increase**. What type of correlation does this represent?

A) Positive correlation

B) Negative correlation

C) No correlation

D) Undefined correlation

Best-Fit Line in Scatter Plots

66. In a scatter plot, the **best-fit line** is used to:

A) Connect all the points

B) Show the overall trend in the data

C) Display the highest and lowest values

D) Change the data values

Section 1: Algebra Foundations

Properties of Real Numbers

67. Which property is illustrated by the equation:

7+(3+2)=(7+3)+2

A) Commutative Property
B) Associative Property
C) Distributive Property
D) Identity Property

Order of Operations (PEMDAS)

68. Solve:

$(6+4)\div2+3\times5$

A) 25
B) 20
C) 29
D) 31

Evaluating Algebraic Expressions

69. If x=2 and y=−3, evaluate:

$4x-y^2+5$

A) 3
B) 7
C) 4
D) 13

Simplifying Expressions

70. Simplify:

$3(2x-5)-4(x+1)$.

A) $2x-11$
B) $2x-19$
C) $6x-11$
D) $6x-19$

Solving Basic Equations & Inequalities

71. Solve:

$5x+3>18$

A) $x>3$
B) $x>5$
C) $x>6$
D) $x>7$

Understanding Absolute Values

72. Solve:

$|x-4|=6$.

A) $x=10,-2$
B) $x=9,-1$
C) $x=8,2$
D) $x=6,-2$

Section 2: Linear Equations & Functions

Understanding Functions (Definition & Notation)

73. Which set of ordered pairs represents a function?

A) (1,2),(2,3),(2,4),(3,5)

B) (3,4),(4,5),(5,6),(6,7)

C) (1,2),(3,4),(5,6),(7,8)

D) (2,3),(3,4),(4,5),(4,6)

Graphing Linear Equations (Slope & Intercepts)

74. What is the slope of the line passing through points (2,3) and (6,7)?

A) 1/2

B) 1

C) 2

D) 3/4

Writing Linear Equations (Slope-Intercept, Point-Slope, & Standard Form)

75. Write the equation of a line that passes through (1,4) with a slope of 3.

A) y=3x+1

B) y=3x+4

C) y=3x+3

D) y=3x+2

Solving Systems of Linear Equations (Graphing, Substitution, & Elimination)

76. Solve using elimination:

3x+2y=10, 5x−2y=6.

A) (4,−1)

B) (2,3)

C) (3,2)

D) (1,4)

Modeling Real-World Problems with Linear Equations

77. A gym charges a $30 monthly fee plus $5 per workout session. Write an equation for the total cost C for x workout sessions.

A) C=30x+5

B) C=5x+30

C) C=30x−5

D) C=5x−30

Section 3: Quadratic Equations & Functions

Understanding Quadratic Functions & Their Graphs

78. Which of the following graphs represents a **quadratic function**?

A) A straight line

B) A U-shaped curve (parabola)

C) A V-shaped graph

D) A stepwise graph

Solving Quadratic Equations (Factoring, Quadratic Formula, Completing the Square)

79. Solve the equation:

$x^2−6x−16=0$.

A) x=8,x=−2

B) x=−8,x=2

C) x=4,x=−4

D) x=6,x=−1

Applications of Quadratics in Real Life

80. A projectile is launched with a height function given by:

$$h(t)=-4t^2+16t+20$$

What is the initial height of the projectile?

A) 4 meters

B) 16 meters

C) 20 meters

D) 36 meters

Quadratic Inequalities

81. Solve the inequality:

$$x^2-9x+14<0.$$

A) 2

B) -2

C) $x<-2$ or $x>7$

D) $x<2$ or $x>7$

Section 4: Exponential Functions & Properties

Exponential Growth & Decay

82. The population of a town is modeled by:

$$P(t)=10,000(1.03)^t$$

What does **1.03** represent?

A) Growth rate of 3%

B) Initial population

C) Doubling time

D) Decay factor

Laws of Exponents

83. Simplify:

$(2x^3y^2)^3$

A) $6x^9y^6$

B) $8x^9y^6$

C) $2x^6y^5$

D) $6x^6y^6$

Simplifying & Solving Exponential Equations

84. Solve for x:

$3^{x+2}=27$

A) x=1

B) x=3

C) x=4

D) x=5

Comparing Linear & Exponential Functions

85. Which function **grows faster** over time?

A) f(x)=5x+7

B) g(x)=1.1^x

C) h(x)=x^2+4

D) j(x)=x^3

Real-World Exponential Function Application

86. A car's value depreciates according to the function:

$$V(t)=30{,}000(0.92)^t$$

What does **0.92** represent?
A) A 92% increase each year
B) A 92% depreciation rate
C) An 8% depreciation rate
D) The car's original price

Section 5: Polynomials & Factoring

Adding, Subtracting, Multiplying, & Dividing Polynomials

87. Simplify:

$$(4x^2+6x-5)-(2x^2-3x+7)$$

A) $2x^2+9x-12$
B) $6x^2+3x+2$
C) $2x^2+9x+2$
D) $2x^2+9x-2$

Factoring Techniques (GCF, Difference of Squares, Trinomials)

88. Factor completely:

$$x^2-10x+25$$

A) $(x-5)(x-5)$
B) $(x+5)(x-5)$
C) $(x-2)(x-3)$
D) $(x-4)(x-6)$

Solving Polynomial Equations

89. Solve for x:

$x^3 - 27 = 0$

A) x=3
B) x=−3
C) x=9
D) x=27

Section 6: Radical & Rational Expressions

Simplifying Square Roots & Cube Roots

90. Simplify:

$\sqrt{98}$

A) $7\sqrt{2}$
B) $2\sqrt{7}$
C) $14\sqrt{2}$
D) $3\sqrt{11}$

Rationalizing Denominators

91. Simplify:

$5/\sqrt{7}$

A) $5\sqrt{7}/7$
B) $7\sqrt{5}/7$
C) $5\sqrt{7}/5$
D) $5\sqrt{7}/14$

Adding, Subtracting, Multiplying, & Dividing Radical Expressions

92. Simplify:

$(5\sqrt{3}+4\sqrt{2})-(2\sqrt{3}+\sqrt{2})$.

A) $3\sqrt{3}+3\sqrt{2}$
B) $3\sqrt{3}+\sqrt{2}$
C) $7\sqrt{3}+5\sqrt{2}$
D) $3\sqrt{3}+4\sqrt{2}$

Section 7: Data Analysis & Probability

Interpreting Graphs & Tables

93. What type of graph is best for **showing parts of a whole**?
A) Bar graph
B) Line graph
C) Pie chart
D) Histogram

Mean, Median, Mode & Range

94. Find the **mean** of the numbers:

$9,15,21,30,45$

A) 22
B) 24
C) 26
D) 28

Probability Rules & Applications

95. A box contains 6 red, 3 blue, and 5 green balls. What is the probability of drawing a **red ball**?

A) 1/2

B) 1/3

C) 6/14

D) 3/14

Scatter Plots & Correlation

96. A scatter plot shows that as **the number of hours spent exercising increases, weight decreases**. What type of correlation does this represent?

A) Positive correlation

B) Negative correlation

C) No correlation

D) Undefined correlation

Section 1: Algebra Foundations

Properties of Real Numbers

97. Which property is illustrated by the equation:

$4 \times (5 \times 2) = (4 \times 5) \times 2$

A) Commutative Property

B) Associative Property

C) Distributive Property

D) Identity Property

Order of Operations (PEMDAS)

98. Solve the expression:

$8 + (3 \times 4) - 6 \div 2$

A) 12
B) 14
C) 17
D) 18

Evaluating Algebraic Expressions

99. If a=3 and b=−4, evaluate:

$2a^2+3b+7$

A) 13
B) 15
C) 17
D) 19

Simplifying Expressions

100. Simplify:

$5(2x−3)−4(x+6)$

A) 6x−39
B) 10x−39
C) 6x−18
D) 10x−18

8.2 ANSWER SHEET - PRACTICE TEST 1

1. Answer: B) a(b+c)=ab+ac

Explanation: The **distributive property** states that **multiplication distributes over addition**, meaning that multiplying a sum is the same as multiplying each addend separately and then adding the products.

2. Answer: B) 16

Explanation: Using **PEMDAS (Parentheses, Exponents, Multiplication/Division, Addition/Subtraction):** $8+4\times(6-2)\div2 = 8+4\times4\div2 = 8+16\div2 = 8+8 = 16$.

3. Answer: C) 43

Explanation:
Substituting x=3 and y=2: $5(3^2)-3(2)+4 = 5(9)-6+4 = 45-6+4=43$.

4. Answer: A) 8x+4

Explanation:
Combine like terms: $3x+5x=8x$, $-2+6=4$

So, the simplified expression is **8x+4**.

5. Answer: B) 4

Explanation:

Rearrange the equation: 5x−2x=9+3, 3x=12=x = 12/3 = 4.

6. Answer: A) x=4,x=2/3

Explanation:

Break into two equations: 3x−7=5 or 3x−7=−5

Solving both: 3x=12 $\Rightarrow$ x=4, 3x=2 $\Rightarrow$ x=2/3.

7. Answer: A) y=x^2

Explanation: A function **assigns exactly one output (y) for each input (x)**. The equation x^2+y^2=4 represents a circle, which is **not a function** because one x-value can have two y-values.

8. Answer: B) 1

Explanation:

Using the **slope formula** m=y2−y1/x2−x1: m=6−2/7−3=4/4=1

9. Answer: A) y=3x−1

Explanation:

Using **y=mx+b**, substitute m=3 x=2, and y=5: 5=3(2)+b, 5=6+b, b=−1.

So, the equation is **y=3x−1.**

10. Answer: C) (1,3)

Explanation:

Substituting y=2x+1 into the second equation: 3x−(2x+1)=5, x−1=5, x=6.

Substituting x=1 into y=2(1)+1: y=3 Solution: **(1, 3)**.

11. Answer: B) C=2x+5

Explanation:

The total cost includes a **fixed fee** of $5 and an additional **$2 per item**:

Total Cost=(Cost per item×Number of items)+Entrance Fee. C=2x+5.

12. Answer: C) A U-shaped curve (parabola)

Explanation:

Quadratic functions have the form $y=ax^2+bx+c$ and their graphs are **parabolas** that open upward or downward.

13. Answer: A) x=3,x=2

Explanation:

Factoring: (x−3)(x−2)=0

Setting each factor to zero: x−3=0⟹x=3, x−2=0⟹x=2.

14. Answer: B) 5 feet

Explanation:

The **initial height** is found by setting t=0:

$h(0)=-16(0)^2+48(0)+5=5$.

15. Answer: A) x<−1 or x>5

Explanation:

Factoring: (x−5)(x+1)>0

Using a sign test, the quadratic is **positive** when x<−1 or x>5.

16. Answer: B) Growth rate

Explanation:

In an exponential growth model $P(t)=P0(1+r)^t$, the **base (1.05) represents growth** at **5% per year**.

17. Answer: A) $2x^2+6x-6$

Explanation:

Distribute the negative and combine like terms:

$(3x^2+2x-1)-(x^2-4x+5) = 3x^2+2x-1-x^2+4x-5 = 2x^2 + 6x - 6$.

18. Answer: A) $5\sqrt{2}$

Explanation:

Factor the square root:

$\sqrt{50}=\sqrt{25\times2}=\sqrt{25}\times\sqrt{2}=5\sqrt{2}$

19. Answer: B) 7.2

Explanation:

$Mean=3+7+7+9+10/5=36/5=7.2$

20. Answer: A) 3/2

Explanation:

Total marbles = 4+3+5=12.
Probability of drawing a **blue marble** = 3/12 **after simplification**.

21. Answer: B) Negative correlation

Explanation:

In a **negative correlation**, as one variable increases, the other decreases. For example, **as the number of study hours decreases, test scores tend to drop.**

22. Answer: A) 5√3/3

Explanation: Multiply numerator and denominator by **√3** to rationalize:

5/√3×√3/√3=5√3/3.

23. Answer: D) √5+√2

Explanation: (2√5+3√2)−(√5+2√2)

Distribute the negative: 2√5−√5+3√2−2√2 = √5+√2

24. Answer: A) x=22

Explanation:
Square both sides to eliminate the square root:

$(\sqrt{x}+3)^2=5^2$ = x+3=25 = x=22.

25. Answer: A) (x−4)(x+4)

Explanation:
Recognizing **difference of squares**: $x^2−16=(x−4)(x+4)$

26. Answer: A) x=2

Explanation:
Rewriting as a **difference of cubes**: $(x−2)(x^2+2x+4)=0$

Setting x−2=0, x=2.

27. Answer: A) x=3

Explanation:
Rewrite 16 as a power of 2: $2^{x+1}=2^4$

Since bases are equal, x+1=4 = x=3.

28. Answer: B) g(x)=3ˣ

Explanation:

Exponential functions **grow at a faster rate** than linear or quadratic functions as xxx increases.

29. Answer: B) Line graph

Explanation:

A **line graph** is used to **display data trends over time**.

30. Answer: B) 15

Explanation:

The **median** is the **middle number** when ordered: 5,8,12,15,22,25,30

Median = **15**.

31. Answer: A) (2×3)×4=2×(3×4)

Explanation:

The **associative property** states that **the way numbers are grouped in multiplication does not change their product**.

32. Answer: C) 19

Explanation: Using **PEMDAS (Parentheses, Exponents, Multiplication/Division, Addition/Subtraction)**: $10+4×3−6÷2 = 10+12−3 = 10 + 12 - 3 = 19$.

33. Answer: B) 1

Explanation:

Substituting a=−2 and b=4:

$2(-2)^2-3(4)+5 = 2(4)-12+5 = 8-12+5 = 1.$

34. Answer: A) 2x+8

Explanation:

Distribute the negative: $4x+5-2x+3$

Combine like terms: $(4x-2x)+(5+3)=2x+8$

35. Answer: B) x=3

Explanation:

Rearrange the equation:

$7x-3x=8+4 = 4x=12 = x=12/4=3.$

36. Answer: A) x=3,-2

Explanation:

Break into two equations: $2x-1=5$ or $2x-1=-5$

Solving both: $2x=6 \Rightarrow x=3$, $2x=-4 \Rightarrow x=-2$

37. Answer: B) $x=y^2-1$

Explanation:

A function assigns **only one output (y) for each input (x)**. The equation $x=y^2-1$ is **not a function** because one x-value can have multiple y-values.

38. Answer: A) (0,-3)

Explanation:

Set x=0 and solve for y: $-2y=-6 = y=3$ So, y-intercept = **(0, -3)**.

39. Answer: B) y=2x-1

Explanation:

Using **y=mx+b**, substitute m=2, x=4, y=7: 7=2(4)+b = b=−1.

Equation: **y=2x−1**

40. Answer: B) y=ax²+bx+c

Explanation:

The **standard form** of a quadratic equation is **y=ax²+bx+c**, where **a, b,** and **c** are constants, and **a≠0**.

41. Answer: A) x=3,x=4

Explanation:

Factoring the quadratic equation: (x−3)(x−4)=0

Setting each factor to zero: x−3=0⟹ x=3, x−4=0⟹x=4.

42. Answer: C) 23 meters

Explanation:

The **maximum height** occurs at the **vertex** of the quadratic function. The vertex formula is:

t=−b/2a, t=−20/2(−5)=2.

Substituting t=2t = 2t=2 into the function: h(2)=−5(2)²+20(2)+3 =−20+40+3=23.

Maximum height = **23 meters**.

43. Answer: B) 2

Explanation:

Factoring: (x−2)(x−3)<0

Solving for the range:

Between **x = 2 and x = 3**, the quadratic is negative. So the solution is: 2

44. Answer: A) Growth rate of 8%

Explanation:

In exponential growth, the formula is: $P(t)=P_0(1+r)^t$

where **r** is the growth rate. Since **1.08=1+0.08**, it means the population grows by **8% per unit of time**.

45. Answer: B) $9x^4y^6$

Explanation:
Applying the **power rule** $(a^m)^n=a^{m\times n}$:

$(3^2)(x^{2\times 2})(y^{3\times 2}) = 9x^4y^6$

46. Answer: B) x=2

Explanation:
Rewriting 125 as a power of 5: $5^{x+1}=5^3$

Since the bases are equal, set exponents equal to each other: x+1=3 = x=2

47. Answer: B) $g(x)=2^x$

Explanation:
Exponential functions **grow faster** than **linear, quadratic, or cubic functions**.

48. Answer: B) A 15% depreciation rate

Explanation:
In exponential decay, the formula is: $V(t)=V_0(1-r)^t$

where **r** is the decay rate. Since **1−0.15=0.85**, the car loses **15% of its value each year**.

49. Answer: C) $7x^2+x+2$

Explanation:

Adding like terms: $(2x^2+5x^2)+(3x−2x)+(−4+6) = 7x^2+x+2$.

50. Answer: A) $(2x−1)(3x−4)$

Explanation:

Using the **AC method**, find two numbers that multiply to **6 × 4 = 24** and add to **-11**:

$(2x−1)(3x−4)=6x^2−11x+4$.

51. Answer: A) x=3

Explanation:

Rewriting as a **difference of cubes**: $(x−3)(x^3+3x+9)=0$

Setting x−3=0, x=3.

52. Answer: A) $5\sqrt{3}$

Explanation:

Factor the square root:

$\sqrt{75}=\sqrt{25×3}=\sqrt{25}×\sqrt{3}=5\sqrt{3}$

53. Answer: A) $7\sqrt{2}/2$

Explanation:

Multiply numerator and denominator by $\sqrt{2}$

$7/\sqrt{2}×\sqrt{2}/\sqrt{2}=7\sqrt{2}/2$.

54. Answer: D) $2\sqrt{3}-\sqrt{5}$

Explanation: $(4\sqrt{3}+2\sqrt{5})-(2\sqrt{3}+\sqrt{5})$

Distribute the negative sign:

$4\sqrt{3}-2\sqrt{3}+2\sqrt{5}-\sqrt{5} = 2\sqrt{3}-\sqrt{5}$.

55. Answer: D) $x=18$

Explanation:
Square both sides to eliminate the square root:

$(\sqrt{x+7})^2=5^2 = x+7=25 = x = 18$.

56. Answer: B) The height of the bars

Explanation:
In a bar graph, the **height of the bars represents the quantity being measured**. The tallest bar indicates the highest sales month.

57. Answer: B) 10

Explanation:
The **mean (average)** is calculated as:

$4+8+10+12+16/5=50/5=10$.

58. Answer: B) 9

Explanation:
Arrange numbers in **ascending order**: $5,7,9,11,15$

Since there is an **odd number of data points**, the median is the **middle value, 9**.

59. Answer: D) 8

Explanation:
The **mode** is the **number that appears most frequently**. Since **8 appears three times**, it is the **mode**.

60. Answer: B) 11

Explanation:
The **range** is the **difference between the highest and lowest values**:

15−4=11.

61. Answer: B) 3/15

Explanation:
Total marbles = 5+3+7=15
Probability of drawing a **blue marble** = 3/15.

62. Answer: C) 28/50

Explanation:
Experimental probability is:

Number of times event occurs/Total trials =28/50.

63. Answer: A) 1/12

Explanation:

- Probability of rolling a **3** = 1/6

- Probability of flipping **tails** = 1/2
  Since these are **independent events**, multiply the probabilities:
  1/6×1/2=1/12.

64. Answer: A) 2/15

Explanation:

- Probability of **first red ball**: 4/12=1/3.

- Probability of **second red ball** (after one red is removed): 3/11.
  Multiply probabilities: 4/12×3/11=12/132=1/11.

65. Answer: A) Positive correlation

Explanation:
A **positive correlation** occurs when **one variable increases as the other also increases**.

66. Answer: B) Show the overall trend in the data

Explanation:
A **best-fit line** helps **identify trends** and **make predictions** about data points.

67. Answer: B) Associative Property

Explanation:
The **Associative Property** states that **the way numbers are grouped in addition or multiplication does not change their sum or product**.

68. Answer: B) 20

Explanation: Using **PEMDAS**: (6+4)÷2+3×5 = 10÷2+15 =5+15=20.

69. Answer: C) 4

Explanation:
Substituting values: $4(2)-(-3)^2+5 = 8-9+5 = 4$.

70. Answer: B) 2x−19

Explanation:

Distribute: $6x-15-4x-4=2x-19$.

71. Answer: B) x>3

Explanation:

$5x>18-3 = 5x>15 = x>3$.

72. Answer: C) x=8,2

Explanation: $x-4=6$ or $x-4=-6$

Solving both: $x=8$ or $x=2$.

73. Answer: C) (1,2),(3,4),(5,6),(7,8)

Explanation:

A **function** assigns **only one output per input**. In choice **A & D**, an input is repeated with different outputs.

74. Answer: B) 1

Explanation:

Using the **slope formula** $m=y2-y1/x2-x1$:

$m=7-3/6-2=4/4=1$.

75. Answer: A) y=3x+1

Explanation:

Using $y=mx+b$, substitute $m=3$ and $(x,y)=(1,4)$: $4=3(1)+b, b=1$

Equation: **y=3x+1**.

76. Answer: C) (3,2)

Explanation:

Adding the equations: $(3x+2y)+(5x-2y)=10+6$, $8x=16$, $x=2$

Substituting $x=2$ into $3x+2y=10$:

$3(2)+2y=10$, $6+2y=10$, $2y=4$, $y=2$.

77. Answer: B) C=5x+30

Explanation:

Total cost includes a **fixed cost of $30** plus **$5 per session**: $C=5x+30$.

78. Answer: B) A U-shaped curve (parabola)

Explanation:

A **quadratic function** follows the form $y=ax^2+bx+c$ and its graph is a **parabola** that can open upwards or downwards.

79. Answer: A) x=8,x=−2

Explanation:

Factoring: $(x-8)(x+2)=0$.

Setting each factor to zero:

$x-8=0 \Rightarrow x=8$, $x+2=0 \Rightarrow x=-2$.

80. Answer: C) 20 meters

Explanation:

The **initial height** is the value of $h(0)$:

$h(0)=-4(0)^2+16(0)+20=20$.

81. Answer: A) 2

Explanation:

Factoring: $(x-2)(x-7)<0$

Between $x=2$ and $x=7$, the quadratic expression is negative. 2

82. Answer: A) Growth rate of 3%

Explanation:

In the exponential growth formula: $P(t)=P_0(1+r)^t$

The base 1.03 means the population **grows by 3% per year**.

83. Answer: B) $8x^9y^6$

Explanation:

Using the **power rule** $(a^m)^n=a^{m\times n}$: $(2^3)(x^{3\times 3})(y^{2\times 3})=8x^9y^6$

84. Answer: A) x=1

Explanation:

Rewrite 27 as a power of 3: $3^{x+2}=3^3$.

Since the bases are equal, set the exponents equal: $x+2=3$ = $x=1$.

85. Answer: B) $g(x)=1.1^x$

Explanation:

Exponential functions grow at a **faster rate** than linear or polynomial functions as x increases.

86. Answer: C) An 8% depreciation rate

Explanation:

In exponential decay, the formula is: $V(t)=V_0(1-r)^t$

Since **$1-0.08=0.92$**, the car loses **8% of its value each year**.

87. Answer: A) $2x^2+9x-12$

Explanation:
Distribute the negative sign and combine like terms:

$4x^2+6x-5-2x^2+3x-7 = (4x^2-2x^2)+(6x+3x)+(-5-7) = 2x^2+9x-12$.

88. Answer: A) $(x-5)(x-5)$

Explanation:
Recognizing **perfect square trinomial**: $x^2-10x+25=(x-5)(x-5)=(x-5)^2$

89. Answer: A) $x=3$

Explanation:
Rewrite as a **difference of cubes**: $(x-3)(x^2+3x+9)=0$

Setting $x-3=0$, $x=3$.

90. Answer: A) $7\sqrt{2}$

Explanation:
Factor the square root: $\sqrt{98}=\sqrt{49\times2}=\sqrt{49}\times\sqrt{2}=7\sqrt{2}$.

91. Answer: A) $5\sqrt{7}/7$

Explanation:
Multiply numerator and denominator by $\sqrt{7}$ to rationalize:

$5/\sqrt{7}\times\sqrt{7}/\sqrt{7}=5\sqrt{7}/7$.

92. Answer: B) $3\sqrt{3}+\sqrt{2}$

Explanation:
Distribute the negative sign:

$(5\sqrt{3}+4\sqrt{2})-(2\sqrt{3}+\sqrt{2}) = (5-2)\sqrt{3}+(4-1)\sqrt{2}=3\sqrt{3}+\sqrt{2}.$

93. Answer: C) Pie chart

Explanation:

A **pie chart** visually represents **parts of a whole** by dividing a circle into proportional sections.

94. Answer: B) 24

Explanation: Mean=9+15+21+30+45/5=120/5=24.

95. Answer: C) 6/14

Explanation:

Total balls = 6+3+5=14
Probability of drawing a **red ball** = 6/14.

96. Answer: B) Negative correlation

Explanation:

A **negative correlation** occurs when **one variable increases as the other decreases**.

97. Answer: B) Associative Property

Explanation:

The **Associative Property** states that **the way numbers are grouped in multiplication does not change their product**.

98. Answer: C) 17

Explanation: Using **PEMDAS**:

$8+(3\times4)-6\div2 = 8+12-3 = 20-3 = 17.$

99. Answer: A) 13

Explanation:

Substituting values: $2(3)^2 + 3(-4) + 7 = 2(9) - 12 + 7 = 18 - 12 + 7 = 13$.

100. Answer: A) 6x−39

Explanation:

Distribute: $10x - 15 - 4x - 24 = 6x - 39$.

9.1 FULL-LENGTH PRACTICE TEST 2

Section 1: Algebra Foundations

Properties of Real Numbers

101. Which of the following properties states that **multiplication distributes over addition**?

A) Associative Property
B) Commutative Property
C) Distributive Property
D) Identity Property

Order of Operations (PEMDAS)

102. Solve:

$$6+2\times(8-3)\div5$$

A) 8
B) 9
C) 10
D) 11

Evaluating Algebraic Expressions

103. If $x=-2$ and $y=5$, evaluate:

$3x^2-4y+7$

A) -2

B) -1

C) 3

D) 5

Simplifying Expressions

104. Simplify:

$(7x-3)+(5x+9)$

A) $12x+6$

B) $12x-6$

C) $2x+6$

D) $2x-12$

Solving Basic Equations & Inequalities

105. Solve:

$4x-5=3x+2$

A) $x=5$

B) $x=7$

C) $x=3$

D) $x=-7$

Understanding Absolute Values

106. Solve for x:

$|x-6|=4$

A) x=10,2

B) x=8,−2

C) x=9,3

D) x=5,−5

Section 2: Linear Equations & Functions

Understanding Functions (Definition & Notation)

107. Which of the following represents a **function**?

A) $y^2=x$

B) $x=y^2+3$

C) $y=2x+5$

D) $x^2+y^2=25$

Graphing Linear Equations (Slope & Intercepts)

108. Find the slope of the line passing through (2,5) and (6,9).

A) 2

B) 4/3

C) 1

D) 3/4

Writing Linear Equations (Slope-Intercept, Point-Slope, & Standard Form)

109. Find the equation of a line that passes through (3,7) with a slope of −2.

A) y=−2x+13

B) y=−2x+7

C) y=−2x−7

D) y=−2x+6

Solving Systems of Linear Equations (Graphing, Substitution, & Elimination)

110. Solve using elimination:

$2x+3y=12$, $4x-3y=6$

A) (3,2)
B) (4,1)
C) (2,4)
D) (1,5)

Modeling Real-World Problems with Linear Equations

111. A movie theater charges $8 per ticket plus a $2 service fee. Write an equation for the total cost C for x tickets.
A) $C=8x+2$
B) $C=8x-2$
C) $C=2x+8$
D) $C=10x$

Section 3: Quadratic Equations & Functions

Understanding Quadratic Functions & Their Graphs

112. Which of the following equations represents a quadratic function?
A) $y=3x-5$
B) $y=2x^2+4x-1$
C) $y=5x^3-2x+7$
D) $y=x+8$

Solving Quadratic Equations (Factoring, Quadratic Formula, Completing the Square)

113. Solve:

$$x^2+7x+12=0$$

A) x=−3,−4
B) x=3,4
C) x=−6,−2
D) x=−2,−5

Applications of Quadratics in Real Life

114. A rocket is launched and its height is given by the function:

$$h(t)=-5t^2+30t+10$$

What is the **maximum height** the rocket reaches?
A) 40 meters
B) 50 meters
C) 55 meters
D) 60 meters

Quadratic Inequalities

115. Solve the inequality:

$$x2-4x-5>0$$

A) x<−1 or x>5
B) −1
C) x<−5 or x>1
D) x<−3 or x>2

Section 4: Exponential Functions & Properties

Exponential Growth & Decay

116. A population of bacteria follows the exponential function:

$P(t)=500(1.05)^t$

What does **1.05** represent?
A) Growth rate of 5%
B) Initial population
C) Decay rate of 5%
D) Half-life

Laws of Exponents

117. Simplify:

$(4x^2y^3)^2$

A) $8x^4y^6$
B) $16x^4y^6$
C) $4x^6y^4$
D) $16x^2y^3$

Simplifying & Solving Exponential Equations

118. Solve for x:

$2^x+^1=16$

A) x=2
B) x=3
C) x=4
D) x=5

Comparing Linear & Exponential Functions

119. Which function grows **faster** as x increases?

A) $f(x)=10x+3$

B) $g(x)=1.2^x$

C) $h(x)=x^2-4x+5$

D) $j(x)=x^3$

Real-World Exponential Function Application

120. A car's value depreciates according to the function:

$$V(t)=20,000(0.88)^t$$

What does **0.88** represent?

A) A 12% increase each year

B) A 12% depreciation rate

C) A 12-year depreciation period

D) The car's initial value

Section 5: Polynomials & Factoring

Adding, Subtracting, Multiplying, & Dividing Polynomials

121. Simplify:

$$(3x^2+2x-5)+(4x^2-6x+8)$$

A) $7x^2-4x+3$

B) $7x^2-4x-13$

C) $7x^2+8x+3$

D) $12x^2-4x+3$

Factoring Techniques (GCF, Difference of Squares, Trinomials)

122. Factor completely:

$x^2-8x+16$

A) $(x-4)(x-4)$
B) $(x+4)(x-4)$
C) $(x-2)(x-8)$
D) $(x-6)(x-10)$

Solving Polynomial Equations

123. Solve for x:

$x^3-64=0$

A) $x=4$
B) $x=-4$
C) $x=8$
D) $x=-8$

Section 6: Radical & Rational Expressions

Simplifying Square Roots & Cube Roots

124. Simplify:

$\sqrt{72}$

A) $6\sqrt{2}$
B) $3\sqrt{8}$
C) $4\sqrt{3}$
D) $2\sqrt{18}$

Rationalizing Denominators

125. Simplify:

$3/\sqrt{5}$

A) $3\sqrt{5}/5$
B) $5\sqrt{3}/5$
C) $3\sqrt{5}/3$
D) $3\sqrt{5}/10$

Solving Radical Equations

126. Solve for x:

$\sqrt{x}+9=7$

A) x=40
B) x=49
C) x=58
D) x=64

Section 7: Data Analysis & Probability

Interpreting Graphs & Tables

127. What type of graph is best for **showing a trend over time**?
A) Bar graph
B) Line graph
C) Pie chart
D) Histogram

Mean, Median, Mode & Range

128. Find the **mean** of the numbers:

14,18,22,26,30

A) 20

B) 22

C) 24

D) 26

Probability Rules & Applications

129. A box contains 3 red, 4 blue, and 5 green balls. What is the probability of drawing a **blue ball**?

A) 2/3

B) 4/12

C) 1/3

D) 5/12

Scatter Plots & Correlation

130. A scatter plot shows that as **hours of study increase, test scores also increase**. What type of correlation does this represent?

A) Positive correlation

B) Negative correlation

C) No correlation

D) Undefined correlation

Section 1: Algebra Foundations

Properties of Real Numbers

131. Which of the following demonstrates the **commutative property of multiplication**?

A) (2+3)+4=2+(3+4)

B) 4×7=7×4

C) 5×(2+3)=(5×2)+(5×3)

D) 8+0=8

Order of Operations (PEMDAS)

132. Solve:

$(8+6) \div 2 + 4 \times 3$

A) 16
B) 18
C) 19
D) 22

Evaluating Algebraic Expressions

133. If a=4 and b=−3, evaluate:

$2a^2 - 3b + 5$

A) 28
B) 35
C) 46
D) 43

Simplifying Expressions

134. Simplify:

$(5x-4) - (3x+7)$

A) 2x−11
B) 2x+3
C) 8x−11
D) 8x+3

Solving Basic Equations & Inequalities

135. Solve:

$6x-3=2x+9$

A) $x=3$
B) $x=4$
C) $x=5$
D) $x=6$

Understanding Absolute Values

136. Solve:

$|2x-5|=7$

A) $x=6,-1$
B) $x=5,-3$
C) $x=4,-2$
D) $x=3,-1$

Section 2: Linear Equations & Functions

Understanding Functions (Definition & Notation)

137. Which set of ordered pairs represents a **function**?
A) $(1,3),(2,4),(2,5),(3,6)$
B) $(4,5),(5,6),(6,7),(7,8)$
C) $(2,3),(3,4),(4,5),(4,6)$
D) $(1,2),(3,4),(5,6),(7,8)$

Graphing Linear Equations (Slope & Intercepts)

138. Find the slope of the line passing through $(1,2)$ and $(5,10)$.
A) $2/3$

B) 3/4

C) 2

D) 4

Writing Linear Equations (Slope-Intercept, Point-Slope, & Standard Form)

139. Find the equation of a line that passes through (2,8) with a slope of 3.

A) y=3x+4

B) y=3x+2

C) y=3x+5

D) y=3x+6

Solving Systems of Linear Equations (Graphing, Substitution, & Elimination)

140. Solve using elimination:

5x+3y=17, 2x−3y=4.

A) (3,2/3)

B) (4,1)

C) (2,5)

D) (1,3)

Section 3: Quadratic Equations & Functions

Understanding Quadratic Functions & Their Graphs

141. Which of the following equations represents a quadratic function?

A) y=2x−5

B) $y=3x^2+4x-7$

C) $y=5x^3-2x+6$

D) y=x+3

Solving Quadratic Equations (Factoring, Quadratic Formula, Completing the Square)

142. Solve the equation:

$$x^2-5x-24=0$$

A) $x=-3,8$
B) $x=-8,3$
C) $x=6,-4$
D) $x=-6,4$

Applications of Quadratics in Real Life

143. A ball is thrown into the air, and its height is modeled by the function:

$$h(t)=-4t^2+16t+20$$

At what time does the ball reach its maximum height?
A) 2 seconds
B) 3 seconds
C) 4 seconds
D) 5 seconds

Quadratic Inequalities

144. Solve the inequality:

$$x^2-2x-15<0$$

A) $x<-3$ or $x>5$
B) -3
C) $x<-5$ or $x>3$
D) $x<3$ or $x>5$

Section 4: Exponential Functions & Properties

Exponential Growth & Decay

145. The value of a car depreciates according to the function:

$$V(t)=30,000(0.85)^t$$

What does **0.85** represent?
A) A 15% growth rate
B) A 15% depreciation rate
C) The initial value of the car
D) The number of years

Laws of Exponents

146. Simplify:

$$(2x^3y^4)^2$$

A) $4x^6y^8$
B) $4x^5y^7$
C) $4x^7y^5$
D) $4x^6y^6$

Simplifying & Solving Exponential Equations

147. Solve for x:

$$3^{x+2}=81$$

A) $x=2$
B) $x=3$
C) $x=4$
D) $x=5$

Comparing Linear & Exponential Functions

148. Which function grows **faster** over time?

A) $f(x)=3x+5$

B) $g(x)=1.5^x$

C) $h(x)=x^2-3x+2$

D) $j(x)=x^3-x$

Real-World Exponential Function Application

149. The population of a city follows the function:

$$P(t)=5000(1.02)^t$$

What does **1.02** represent?

A) A 2% annual decrease

B) A 2% annual growth rate

C) The initial population

D) The decay factor

Section 5: Polynomials & Factoring

Adding, Subtracting, Multiplying, & Dividing Polynomials

150. Simplify:

$$(4x^2-2x+7)+(3x^2+5x-4)$$

A) $7x^2+3x+3$

B) $7x^2-3x-3$

C) $7x^2+3x-3$

D) $7x^2+7x+11$

Factoring Techniques (GCF, Difference of Squares, Trinomials)

151. Factor completely:

$x^2+9x+20$

A) $(x+4)(x+5)$
B) $(x-4)(x-5)$
C) $(x+2)(x+10)$
D) $(x-2)(x-10)$

Solving Polynomial Equations

152. Solve for x:

$x^3-27=0$

A) $x=3$
B) $x=-3$
C) $x=9$
D) $x=27$

Section 6: Radical & Rational Expressions

Simplifying Square Roots & Cube Roots

153. Simplify:

$\sqrt{50}$

A) $5\sqrt{2}$
B) $10\sqrt{5}$
C) $3\sqrt{10}$
D) $4\sqrt{2}$

Rationalizing Denominators

154. Simplify:

$6/\sqrt{3}$

A) $6\sqrt{3}/3$
B) $2\sqrt{3}/3$
C) $6\sqrt{3}/9$
D) $3\sqrt{6}/2$

Solving Radical Equations

155. Solve for x:

$\sqrt{x}+4=6$

A) x=32
B) x=36
C) x=40
D) x=42

Section 7: Data Analysis & Probability

Interpreting Graphs & Tables

156. What type of graph is best for **showing categories of data**?
A) Bar graph
B) Line graph
C) Pie chart
D) Histogram

Mean, Median, Mode & Range

157. Find the **mean** of the numbers:

14,16,18,20,22,24

A) 16

B) 17

C) 18

D) 19

Probability Rules & Applications

158. A box contains 4 red, 5 blue, and 6 green balls. What is the probability of drawing a **blue ball**?

A) 5/15

B) 1/15

C) 4/15

D) 5/14

Scatter Plots & Correlation

159. A scatter plot shows that as **exercise increases, weight decreases**. What type of correlation does this represent?

A) Positive correlation

B) Negative correlation

C) No correlation

D) Undefined correlation

Best-Fit Line in Scatter Plots

160. A best-fit line on a scatter plot is used to:

A) Connect all data points

B) Predict future values

C) Represent the mean of all points

D) Display the highest and lowest values

Section 1: Algebra Foundations

Properties of Real Numbers

161. Which property is illustrated by the equation:

$(a+b)+c=a+(b+c)$

A) Commutative Property
B) Associative Property
C) Distributive Property
D) Identity Property

Order of Operations (PEMDAS)

162. Solve:

$8+2\times(6-4)\div2$

A) 10
B) 12
C) 14
D) 16

Evaluating Algebraic Expressions

163. If a=3 and b=−2, evaluate:

$4a^2-5b+7$

A) 44
B) 48
C) 51
D) 53

Simplifying Expressions

164. Simplify:

$(6x-5)-(2x+4)$

A) $4x-9$
B) $4x+9$
C) $8x-1$
D) $8x+1$

Solving Basic Equations & Inequalities

165. Solve:

$7x+3=4x+12$

A) $x=1$
B) $x=2$
C) $x=3$
D) $x=4$

Understanding Absolute Values

166. Solve:

$|x-3|=7$

A) $x=10,-4$
B) $x=9,-3$
C) $x=8,-2$
D) $x=7,-1$

Section 2: Linear Equations & Functions

Understanding Functions (Definition & Notation)

167. Which of the following sets represents a function?

A) (1,2),(2,3),(2,4),(3,5)

B) (4,5),(5,6),(6,7),(7,8)

C) (2,3),(3,4),(4,5),(4,6)

D) (1,2),(3,4),(5,6),(7,8)

Graphing Linear Equations (Slope & Intercepts)

168. Find the slope of the line passing through (3,7) and (6,13).

A) 23

B) 34

C) 2

D) 4

Writing Linear Equations (Slope-Intercept, Point-Slope, & Standard Form)

169. Find the equation of a line that passes through (1,4) with a slope of -3.

A) y=−3x+7

B) y=−3x+4

C) y=−3x+1

D) y=−3x+6

Solving Systems of Linear Equations (Graphing, Substitution, & Elimination)

170. Solve using elimination:

3x+2y=14, 5x−2y=6.

A) (4,1)

B) (3,2)

C) (2,4)

D) (1,5)

Section 3: Quadratic Equations & Functions

Understanding Quadratic Functions & Their Graphs

171. Which function represents a **quadratic equation**?

A) $y=3x+4$

B) $y=x^2-6x+8$

C) $y=5x^3-2x+1$

D) $y=4/x+2$

Solving Quadratic Equations (Factoring, Quadratic Formula, Completing the Square)

172. Solve for x:

$x^2+5x-24=0$

A) $x=3,-8$

B) $x=4,-6$

C) $x=-3,8$

D) $x=6,-4$

Applications of Quadratics in Real Life

173. A stone is thrown into the air, and its height in feet is modeled by:

$h(t)=-16t^2+32t+5$

At what time does the stone reach its maximum height?

A) 0.5 seconds

B) 1 second

C) 2 seconds

D) 3 seconds

Quadratic Inequalities

174. Solve the inequality:

$x^2-7x+10<0$

A) $x<-2$ or $x>5$

B) -2

C) 2

D) $x<3$ or $x>5$

Section 4: Exponential Functions & Properties

Exponential Growth & Decay

175. A city's population follows the function:

$P(t)=10,000(1.04)^t$

What does **1.04** represent?

A) A 4% annual decrease

B) A 4% annual growth rate

C) The initial population

D) The decay factor

Laws of Exponents

176. Simplify:

$(3x^2y^4)^3$

A) $9x^6y^{12}$

B) $27x^6y^{12}$

C) $9x^5y^9$

D) $27x^5y^{10}$

Simplifying & Solving Exponential Equations

177. Solve for x:

$2^{x+1}=32$

A) x=3

B) x=4

C) x=5

D) x=6

Comparing Linear & Exponential Functions

178. Which function grows **faster** over time?

A) $f(x)=4x+2$

B) $g(x)=1.3^x$

C) $h(x)=x^2-5x+3$

D) $j(x)=x^3-x$

Real-World Exponential Function Application

179. The value of a car depreciates according to the function:

$V(t)=25,000(0.90)^t$

What does **0.90** represent?

A) A 10% increase each year

B) A 10% depreciation rate

C) The car's initial price

D) The time in years

Section 5: Polynomials & Factoring

Adding, Subtracting, Multiplying, & Dividing Polynomials

180. Simplify:

$$(5x^2+3x-7)+(2x^2-4x+9)$$

A) $3x^2-x+16$

B) $7x^2-x+2$

C) $7x^2-x+16$

D) $3x^2+x-2$

Factoring Techniques (GCF, Difference of Squares, Trinomials)

181. Factor completely:

$$x^2-11x+30$$

A) $(x-5)(x-6)$

B) $(x-3)(x-10)$

C) $(x-2)(x-15)$

D) $(x-4)(x-8)$

Solving Polynomial Equations

182. Solve for x:

$$x^3-8=0$$

A) $x=2$

B) $x=-2$

C) x=4

D) x=−4

Section 6: Radical & Rational Expressions

Simplifying Square Roots & Cube Roots

183. Simplify:

$\sqrt{98}$

A) $7\sqrt{2}$

B) $9\sqrt{3}$

C) $5\sqrt{4}$

D) $3\sqrt{8}$

Rationalizing Denominators

184. Simplify:

$4/\sqrt{5}$

A) $4\sqrt{5}/5$

B) $5\sqrt{4}/4$

C) $3\sqrt{5}/5$

D) $2\sqrt{5}/5$

Solving Radical Equations

185. Solve for x:

$\sqrt{x}+9=5$

A) x=14

B) x=16

C) x=20

D) x=25

Section 7: Data Analysis & Probability

Interpreting Graphs & Tables

186. What type of graph is best for **showing trends over time**?

A) Bar graph

B) Line graph

C) Pie chart

D) Histogram

Mean, Median, Mode & Range

187. Find the **median** of the numbers:

7,12,14,21,25

A) 12

B) 14

C) 16

D) 21

Probability Rules & Applications

188. A bag contains 3 red, 5 blue, and 7 green marbles. What is the probability of drawing a **blue marble**?

A) 5/15

B) 1/15

C) 5/14

D) 3/14

Scatter Plots & Correlation

189. A scatter plot shows that as **studying time increases, exam scores increase**. What type of correlation does this represent?

A) Positive correlation

B) Negative correlation

C) No correlation

D) Undefined correlation

Best-Fit Line in Scatter Plots

190. What is the purpose of a best-fit line in a scatter plot?

A) To connect all data points

B) To predict future values

C) To represent the average of all points

D) To display the highest and lowest values

Section 1: Algebra Foundations

Properties of Real Numbers

191. Which property is used in the equation:

$4(x+3)=4x+12$

A) Associative Property

B) Commutative Property

C) Distributive Property

D) Identity Property

Order of Operations (PEMDAS)

192. Solve:

$10-3\times4+6\div2$

A) 1

B) 4

C) 6

D) 8

Evaluating Algebraic Expressions

193. If x=−2 and y=3, evaluate:

$5x^2−4y+8$

A) 10

B) 12

C) 14

D) 16

Simplifying Expressions

194. Simplify:

(8x−6)−(3x+7)

A) 5x−13

B) 5x+1

C) 11x−13

D) 11x+1

Solving Basic Equations & Inequalities

195. Solve:

2x−4=3x+1

A) x=−5

B) x=−3

C) x=5

D) x=3

Understanding Absolute Values

196. Solve:

$|x+2|=6$

A) x=4,−8

B) x=3,−9

C) x=2,−10

D) x=5,−7

Section 2: Linear Equations & Functions

Understanding Functions (Definition & Notation)

197. Which of the following sets represents a function?

A) (1,2),(2,3),(3,2),(3,4)

B) (4,5),(5,6),(6,7),(7,8)

C) (2,3),(3,4),(4,5),(4,6)

D) (1,2),(3,4),(5,6),(7,8)

Graphing Linear Equations (Slope & Intercepts)

198. Find the slope of the line passing through (2,5) and (6,13).

A) 2/3

B) 3/4

C) 2

D) 4

Writing Linear Equations (Slope-Intercept, Point-Slope, & Standard Form)

199. Find the equation of a line that passes through (2,5) with a slope of -3.

A) $y=-3x+8$

B) $y=-3x+5$

C) $y=-3x+3$

D) $y=-3x+11$

Solving Systems of Linear Equations (Graphing, Substitution, & Elimination)

200. Solve using elimination:

$4x+3y=18$, $6x-3y=6$.

A) $(2.4, 2.8)$

B) $(3,2)$

C) $(4,1)$

D) $(5,0)$

9.2 ANSWER SHEET – PRACTICE TEST 2

101. Answer: C) Distributive Property

Explanation: The **distributive property** states that multiplying a sum by a number is the same as multiplying each addend separately and then adding the products. $a(b+c)=ab+ac$

102. Answer: A) 8

Explanation: Using **PEMDAS**: $6+2\times(8-3)\div5 = 6+2\times5\div5 = 6+2\times1 = 6+2 = 8$.

103. Answer: B) -1

Explanation: $3(-2)2-4(5)+7 = 3(4)-20+7 = 12-20+7 = -1$.

104. Answer: A) 12x+6

Explanation: $(7x+5x)+(-3+9)=12x+6$

105. Answer: B) x=7

Explanation: $4x-3x=2+5$, $x=7$.

106. Answer: A) x=10,2

Explanation: Break into two equations: $x-6=4$ or $x-6=-4$, $x=10$ or $x=2$.

107. Answer: C) y=2x+5

Explanation:

A function has **one unique output for each input**. The equation $y=2x+5$ follows this rule.

108. Answer: C) 1

Explanation: Using the **slope formula** $m = y2 - y1/x2 - x1$: $m = 9 - 5/6 - 2 = 4/4 = 1$.

109. Answer: A) y=−2x+13

Explanation: Using $y=mx+b$, substitute $m=-2$ and $(x,y)=(3,7)$: $7=-2(3)+b$, $b=13$

Equation: **y=−2x+13**.

110. Answer: A) (3,2)

Explanation: Adding both equations: $(2x+3y)+(4x-3y)=12+6$, $6x=18$, $x=3$.

Substituting $x=3$ into $2x+3y=12$:

$2(3)+3y=12$, $6+3y=12$, $3y=6$ $y=2$. Solution: **(3,2)**.

111. Answer: A) C=8x+2

Explanation: Total cost includes a **\$2 fixed fee** plus **\$8 per ticket**: $C=8x+2$.

112. Answer: B) y=2x²+4x−1

Explanation: A **quadratic function** has the form $y=ax^2+bx+c$, where $a \neq 0$. The only equation in this form is **y=2x²+4x−1**.

113. Answer: A) x=−3,−4

Explanation: Factoring: $(x+3)(x+4)=0$

Setting each factor to zero:

x+3=0⟹x=−3, x+4=0⟹x=−4.

114. Answer: C) 55 meters

Explanation: The **maximum height** occurs at the **vertex** of the quadratic function. The vertex formula is: $t=-b/2a$, $t=-30/2(-5)=3$.

Substituting $t=3$ into the function:

$h(3)=-5(3)^2+30(3)+10 =-45+90+10=55$.

115. Answer: A) x<−1 or x>5

Explanation: Factoring: $(x-5)(x+1)>0$

The quadratic expression is positive when **x<−1 or x>5.**

116. Answer: A) Growth rate of 5%

Explanation: In the exponential growth formula: $P(t)=P_0(1+r)^t$

Since **1.05=1+0.05**, it represents a **5% growth rate** per unit of time.

117. Answer: B) $16x^4y^6$

Explanation: Applying the **power rule** $(a^m)^n=a^{m \times n}$:

$(4^2)(x^2 \times ^2)(y^3 \times ^2)=16x^4y^6$

118. Answer: B) x=3

Explanation: Rewriting 16 as a power of 2: $2^{x+1}=2^4$

Since the bases are equal, set the exponents equal to each other: $x+1=4$, $x=3$.

119. Answer: B) $g(x)=1.2^x$

Explanation:

Exponential functions grow at a **faster rate** than linear or polynomial functions for large values of x.

120. Answer: B) A 12% depreciation rate

Explanation: In exponential decay, the formula is: $V(t)=V_0(1-r)^t$

Since **1−0.12=0.88**, the car loses **12% of its value each year**.

121. Answer: A) $7x^2-4x+3$

Explanation: Adding like terms:

$(3x^2+4x^2)+(2x-6x)+(-5+8)=7x^2-4x+3$.

122. Answer: A) (x−4)(x−4)

Explanation: Recognizing **perfect square trinomial**: $x^2-8x+16=(x-4)(x-4)=(x-4)^2$

123. Answer: A) x=4

Explanation: Rewriting as a **difference of cubes**: $(x-4)(x^2+4x+16)=0$.

Setting x−4=0, x=4.

124. Answer: A) 6√2

Explanation: Factor the square root:

$\sqrt{72}=\sqrt{36\times2}=\sqrt{36}\times\sqrt{2}=6\sqrt{2}$.

125. Answer: A) 3√5/5

Explanation: Multiply numerator and denominator by **√5**:

$3/\sqrt{5} \times \sqrt{5}/\sqrt{5} = 3\sqrt{5}/5$.

126. Answer: A) x=40

Explanation: Square both sides to eliminate the square root:

$(\sqrt{(x+9)})^2 = 7^2$, $x+9 = 49$, $x = 40$.

127. Answer: B) Line graph

Explanation: A **line graph** is used to display **trends** over time.

128. Answer: B) 22

Explanation: Mean$=14+18+22+26+30/5=110/5=22$.

129. Answer: C) 1/3

Explanation: Total balls $= 3+4+5=12$.
Probability of drawing a **blue ball** $= 4/12=1/3$.

130. Answer: A) Positive correlation

Explanation: A **positive correlation** occurs when **one variable increases as the other also increases**.

131. Answer: B) 4×7=7×4

Explanation: The **commutative property of multiplication** states that changing the order of factors does not change the product.

132. Answer: C) 19

Explanation: Using **PEMDAS**:

$(8+6) \div 2+4 \times 3 = 14 \div 2+4 \times 3 = 7+12=19$.

133. Answer: C) 46

Explanation: $2(4)^2 - 3(-3) + 5 = 2(16) + 9 + 5 = 32 + 9 + 5 = 46$.

134. Answer: A) 2x−11

Explanation: $(5x-4)-(3x+7) = 5x-4-3x-7 = (5x-3x)+(-4-7) = 2x-11$.

135. Answer: C) x=3

Explanation: $6x-2x=9+3$, $4x=12$, $x=3$.

136. Answer: A) x=6,−1

Explanation: Break into two equations: $2x-5=7$ or $2x-5=-7$

Solving both: $2x=12 \Rightarrow x=6$, $2x=-2 \Rightarrow x=-1$.

137. Answer: D) (1,2),(3,4),(5,6),(7,8)

Explanation: A **function** assigns **one unique output per input**. In choices **A & C**, an input is repeated with different outputs.

138. Answer: C) 2

Explanation: Using the **slope formula** $m = y_2 - y_1/x_2 - x_1$: $m = 10-2/5-1 = 8/4 = 2$.

139. Answer: A) y=3x+4

Explanation: Using $y=mx+b$, substitute $m=3$ and $(x,y)=(2,8)$: $8=3(2)+b$, $b=8-6=4$.

Equation: **y=3x+4.**

140. Answer: A) (3,2/3)

Explanation: Adding both equations: $(5x+3y)+(2x-3y)=17+4$, $7x=21$, $x=3$.

Substituting $x=3$ into $5x+3y=17$:

$5(3)+3y=17$, $15+3y=17$, $3y=2 \Rightarrow y=2/3$.

Solution: **(3,2/3)**.

141. Answer: B) $y=3x^2+4x-7$

Explanation: A **quadratic function** follows the form $y=ax^2+bx+c$, where $a\neq0$. The only equation that follows this pattern is **$y=3x^2+4x-7$**.

142. Answer: A) $x=-3,8$

Explanation: Factoring the quadratic equation: $(x+3)(x-8)=0$

Setting each factor to zero: $x+3=0 \Rightarrow x=-3$, $x-8=0 \Rightarrow x=8$.

143. Answer: A) 2 seconds

Explanation: The **maximum height** occurs at the **vertex** of the quadratic function. The vertex formula is: $t=-b/2a$, $t=-16/2(-4)=2$.

144. Answer: B) -3

Explanation: Factoring: $(x-5)(x+3)<0$

The quadratic expression is negative between **$x=-3$ and $x=5$.**

145. Answer: B) A 15% depreciation rate

Explanation: The **base of the exponent** represents $1-r$, where rrr is the decay rate. Since $0.85=1-0.15$, the car loses **15% of its value per year**.

146. Answer: A) $4x^6y^8$

Explanation: Applying the **power rule** $(a^m)^n = a^{m \times n}$:

$$(2^2)(x^{3 \times 2})(y^{4 \times 2}) = 4x^6 y^8.$$

147. Answer: A) x=2

Explanation: Rewriting 81 as a power of 3: $3^{x+2} = 3^4$

Since the bases are equal, set the exponents equal to each other: x+2=4, x=2.

148. Answer: B) $g(x)=1.5^x$

Explanation: Exponential functions grow at a **faster rate** than linear or polynomial functions as xxx increases.

149. Answer: B) A 2% annual growth rate

Explanation: In exponential growth, the formula is: $P(t) = P_0(1+r)^t$

Since **1.02=1+0.02**, it represents a **2% annual growth rate**.

150. Answer: A) 7x2+3x+3

Explanation: Adding like terms: $(4x^2 + 3x^2) + (-2x + 5x) + (7-4) = 7x^2 + 3x + 3$.

151. Answer: A) (x+4)(x+5)

Explanation: Two numbers that multiply to **20** and add to **9** are **4 and 5**:

$(x+4)(x+5) = x^2 + 9x + 20$

152. Answer: A) x=3

Explanation: Rewriting as a **difference of cubes**: $(x-3)(x^2 + 3x + 9) = 0$

Setting x−3=0, x=3.

153. Answer: A) 5√2

Explanation: Factor the square root: √50=√25×2=√25×√2=5√2.

154. Answer: A) 6√3/3

Explanation: Multiply numerator and denominator by √3 to rationalize: 6/√3×√3/√3=6√3/3.

155. Answer: C) x=32

Explanation: Square both sides to eliminate the square root: √((x+4))²=6², x+4=36, x=32.

156. Answer: A) Bar graph

Explanation: A **bar graph** is used to compare different categories of data.

157. Answer: D) 19

Explanation: Mean=14+16+18+20+22+24/6=114/6=19.

158. Answer: A) 5/15

Explanation: Total balls = 4+5+6=15.
Probability of drawing a **blue ball** = 5/15.

159. Answer: B) Negative correlation

Explanation: A **negative correlation** occurs when **one variable increases as the other decreases**.

160. Answer: B) Predict future values

Explanation: A **best-fit line** helps **identify trends** and **make predictions** about data points.

161. Answer: B) Associative Property

Explanation: The **Associative Property** states that **changing the grouping of numbers in addition or multiplication does not change the result**.

162. Answer: A) 10

Explanation: Using **PEMDAS**:

$8+2\times(6-4)\div2 = 8+2\times2\div2 = 8+4\div2= 8+2 = 10$.

163. Answer: D) 53

Explanation: $4(3)^2-5(-2)+7 = 4(9)+10+7 = 36+10+7 = 53$.

164. Answer: A) 4x−9

Explanation: $(6x-5)-(2x+4) = 6x-5-2x-4=4x-9$.

165. Answer: C) x=3

Explanation: $7x-4x=12-3$, $3x=9$, $x=3$.

166. Answer: A) x=10,−4

Explanation: Break into two equations: $x-3=7$ or $x-3=-7$

Solving both: $x=10$, $x=-4$.

167. Answer: D) (1,2),(3,4),(5,6),(7,8)

Explanation: A **function** assigns **one unique output per input**. In choices **A & C**, an input is repeated with different outputs.

168. Answer: C) 2

Explanation: Using the **slope formula** $m=y2-y1/x2-x1$:

m=13−7/6−3=6/3=2.

169. Answer: A) y=−3x+7

Explanation: Using y=mx+b, substitute m=−3 and (x,y)=(1,4: 4=−3(1)+b, b=4+3=7.

Equation: **y=−3x+7.**

170. Answer: A) (4,1)

Explanation: Adding both equations:

(3x+2y)+(5x−2y)=14+6, 8x=20, x=4.

Substituting x=4 into 3x+2y=14:

3(4)+2y=14, 12+2y=14, 2y=2, y=1.

171. Answer: B) y=x²−6x+8

Explanation: A **quadratic function** has the form y=ax²+bx+c, where a≠0. The correct choice follows this pattern.

172. Answer: D) x=6,−4

Explanation: Factoring the quadratic equation: (x−6)(x+4)=0

Setting each factor to zero: x−6=0⟹x=6, x+4=0⟹x=−4.

173. Answer: B) 1 second

Explanation: The **maximum height** occurs at the **vertex** of the quadratic function. The vertex formula is: t= −b/2a, t =−32/2(−16) = 1.

174. Answer: C) 2

Explanation: Factoring: $(x-2)(x-5)<0$

The quadratic expression is negative between **x=2 and x=5**.

175. Answer: B) A 4% annual growth rate

Explanation:
In exponential growth, the formula is: $P(t)=P_0(1+r)^t$

Since **1.04=1+0.04**, it represents a **4% annual growth rate**.

176. Answer: B) $27x^6y^{12}$

Explanation: Applying the **power rule** $(a^m)^n=a^{m\times n}$: $(3^3)(x^{2\times3})(y^{4\times3})=27x^6y^{12}$.

177. Answer: B) x=4

Explanation: Rewriting 32 as a power of 2: $2^{x+1}=2^5$

Since the bases are equal, set the exponents equal to each other: $x+1=5$, $x=4$.

178. Answer: B) $g(x)=1.3^x$

Explanation: Exponential functions grow at a **faster rate** than linear or polynomial functions as x increases.

179. Answer: B) A 10% depreciation rate

Explanation: In exponential decay, the formula is: $V(t)=V_0(1-r)^t$

Since **1−0.10=0.90**, the car loses **10% of its value per year**.

180. Answer: B) $7x^2-x+2$

Explanation: Adding like terms: $(5x^2+2x^2)+(3x-4x)+(-7+9)=7x^2-x+2$.

181. Answer: A) (x−5)(x−6)

Explanation: Two numbers that multiply to **30** and add to **-11** are **-5 and -6**:

$(x-5)(x-6)=x^2-11x+30$.

182. Answer: A) x=2

Explanation: Rewriting as a **difference of cubes**: $(x-2)(x^2+2x+4)=0$

Setting $x-2=0$, $x=2$.

183. Answer: A) 7√2

Explanation: Factor the square root: $\sqrt{98}=\sqrt{49\times2}=\sqrt{49}\times\sqrt{2}=7\sqrt{2}$

184. Answer: A) 4√5/5

Explanation: Multiply numerator and denominator by **√5**:

$4/\sqrt{5}\times\sqrt{5}/\sqrt{5}=4\sqrt{5}/5$

185. Answer: B) x=16

Explanation: Square both sides to eliminate the square root:

$\sqrt{(x+9)^2}=5^2$, $x+9=25$, $x=16$.

186. Answer: B) Line graph

Explanation: A **line graph** is used to display **trends** over time.

187. Answer: B) 14

Explanation: The **median** is the middle number when arranged in order: 7,12,14,21,25

The middle value is **14**.

188. Answer: A) 5/15

Explanation: Total marbles = 3+5+7=15.
Probability of drawing a **blue marble** = 5/15.

189. Answer: A) Positive correlation

Explanation: A **positive correlation** occurs when **one variable increases as the other also increases**.

190. Answer: B) To predict future values

Explanation: A **best-fit line** helps **identify trends** and **make predictions** about data points.

191. Answer: C) Distributive Property

Explanation: The **distributive property** states that multiplying a sum by a number is the same as multiplying each term separately: $a(b+c)=ab+ac$

192. Answer: A) 1

Explanation: Using **PEMDAS**:

$10-3\times4+6\div2 =10-12+3= 1$.

193. Answer: D) 16

Explanation: $5(-2)2-4(3)+8 =5(4)-12+8 =20-12+8=16$.

194. Answer: A) 5x−13

Explanation: $(8x-6)-(3x+7) = 8x - 6 - 3x - 7 = (8x - 3x) + (-6 - 7) = 5x - 13$.

195. Answer: A) x=−5

Explanation: $2x-3x=1+4$, $x=-5$.

196. Answer: A) x=4,−8

Explanation: Break into two equations: x+2=6 or x+2=−6.

Solving both: x=4, x=−8.

197. Answer: D) (1,2),(3,4),(5,6),(7,8)

Explanation: A **function** assigns **one unique output per input**. In choices **A & C**, an input is repeated with different outputs.

198. Answer: C) 2

Explanation: Using the **slope formula** m=y2−y1/x2−x1 : m=13−5/6−2=8/4=2.

199. Answer: D) y=−3x+11

Explanation: Using y=mx+b, substitute m=−3 and (x,y)=(2,5): 5=−3(2)+b, b=5+6=11.

Equation: **y=−3x+11.**

200. Answer: A) (2.4,2.8)

Explanation: Adding both equations: (4x+3y)+(6x−3y)=18+6, 10x=24, x=2.4.

Substituting x=2.4 into 4x+3y=18: 4(2.4)+3y=18, 9.6+3y=18, 3y=8.4 y=8.4/3, y=2.8

TEST-TAKING STRATEGIES

Preparing for the **Florida Algebra 1 End-of-Course (EOC) Exam** is more than just studying formulas and solving equations—it's about developing **smart test-taking strategies** and **managing stress effectively**. This section provides **expert strategies** to help you approach the test with **confidence, efficiency, and focus**.

Part 1: Test-Taking Strategies for the Algebra 1 EOC Exam

Understand the Exam Format

- Familiarize yourself with **question types**: Multiple-choice, grid-in, multi-select, and drag-and-drop.

- Know the **weight of each section** and focus on high-yield topics.

- Be aware that **Session 1 does not allow calculators**, so strengthen your mental math skills.

Manage Your Time Wisely

Time Allocation Strategy:

- You have **160 minutes** for about **60-66 questions** (~2.5 minutes per question).

- **Don't spend too much time on one question!** If stuck, mark it and move on.

- Allocate at least **5−10 minutes at the end** to review flagged questions.

Read the Question Carefully

- Look for **keywords** like *"simplify," "solve for x," "evaluate," or "factor."*

- **Underline important numbers and expressions** before solving.

- Watch out for **tricky wording** (e.g., "Which of the following is NOT true?").

Use the Process of Elimination

Strategy:

- **Cross out incorrect answer choices** to narrow options.

- If unsure, **eliminate at least two wrong choices** before guessing.

- For **grid-in responses**, double-check your work before entering your final answer.

Check Your Work Efficiently

- If you have time left, **review flagged questions first**.

- **Recalculate tricky problems** to confirm your answers.

- **For algebraic expressions**, plug in values to verify solutions.

Be Strategic with the Calculator (Session 2 Only)

- **Know when to use the calculator**—some problems are faster to solve mentally.

- Use the **calculator to double-check answers** on complex calculations.

- Avoid careless mistakes by checking **parentheses placement and decimal accuracy**.

Handling Grid-In and Multi-Select Questions

Grid-In Questions:

- Be **precise**—enter exact values (fractions, decimals, or whole numbers).

- **Do not round unless instructed!**

- If needed, **simplify fractions** before entering.

Multi-Select Questions:

- More than one answer may be correct.

- **Select all valid choices**, even if unsure about one or two.

- **Partial credit is not given**, so double-check each selection.

Part 2: Overcoming Test Anxiety

Prepare Early to Build Confidence

- Follow a **structured study plan** (1-month, 2-month, or 3-month schedule).

- Take **timed practice tests** to simulate the real exam environment.

- Get familiar with the **exam format** to reduce surprises on test day.

Stay Positive & Use Motivational Self-Talk

Replace negative thoughts with positive affirmations:

- Instead of **"I can't do this,"** say **"I've studied well, and I'm prepared."**

- Instead of **"I always fail math tests,"** say **"I've practiced, and I know what to expect."**

- Instead of **"I'll never pass,"** say **"I've overcome challenges before—I** can do this!"

Practice Relaxation Techniques

Before & During the Test:

- Take **deep breaths**: Inhale for **4 seconds**, hold for **4 seconds**, exhale for **6 seconds**.

- **Stretch your arms and legs** before the test starts to release tension.

- Use **progressive muscle relaxation**: Tense and release your muscles one at a time.

Use the "Brain Dump" Method at the Start of the Test

How it Works:

- As soon as the test begins, **write down key formulas, rules, and concepts** on your scratch paper.

- This helps **reduce memory overload** and provides a quick reference during the test.

Handle Difficult Questions with a Plan

Stuck on a question? Use these strategies:

Mark it and move on—don't waste time if unsure.

Eliminate wrong answers—guess wisely if needed.

Look for patterns—sometimes plugging in numbers helps.

Break it into smaller steps—simplify complicated expressions.

Sleep & Nutrition Matter!

Night Before the Exam:

- Get **7–9 hours of sleep** for better focus.

- Avoid **last-minute cramming**—a brief review is fine, but don't stress yourself.

Healthy Brain Fuel on Test Day:

- Eat a **balanced breakfast** with protein (eggs, yogurt, nuts) and complex carbs (oatmeal, whole grains).

- Avoid **sugary foods**—they cause energy crashes!

Stay Calm & Focused During the Exam

- If you feel overwhelmed, **pause, take a deep breath, and refocus.**

- **Remind yourself:** "I've studied well. I'm prepared for this."

- Use **positive visualization**—imagine yourself solving problems with confidence.

Part 3: Exam Day Plan & Checklist

One Week Before the Exam

✔ Complete a **full-length practice test** under timed conditions.

✔ Review **formulas, common mistakes, and problem-solving strategies**.

✔ Organize all necessary materials (ID, pencils, calculator).

Day Before the Exam

✔ **Get enough rest** (7–9 hours of sleep).

✔ **Lightly review notes**—avoid heavy cramming.

✔ Pack your **test-day essentials**:

- Photo ID

- Pencils & erasers

- Scientific calculator (for Session 2)

- Scratch paper (if allowed)

On Exam Day

✔ Eat a **healthy breakfast** for sustained energy.

✔ **Arrive early** at the test center to avoid stress.

✔ **Stay positive and focused**—trust your preparation!

Final Words

The **Florida Algebra 1 EOC Exam** is an important milestone, but **you are fully capable of succeeding**. By using **smart test-taking strategies**, practicing effectively, and managing anxiety, you can **boost your confidence and maximize your score**.

ADDITIONAL RESOURCES

RECOMMENDED ONLINE RESOURCES AND ACADEMIC MATERIALS

To **excel** in the **Florida Algebra 1 End-of-Course (EOC) Exam**, it is essential to use **high-quality study materials** and **reliable online resources**. This section provides a curated list of **trusted websites, practice tools, and textbooks** to help you **deepen your understanding, practice effectively, and boost your confidence** before the exam.

Recommended Online Resources

Florida Department of Education (FLDOE) – Official EOC Resources

https://www.fldoe.org
Why Use It?
Official test specifications, sample questions, and EOC practice materials.
Information on exam structure, scoring, and calculator policies.
Access to updated **Florida Standards Assessments (FSA)** guidelines.

Algebra Nation (Florida-Specific Algebra Help)

https://www.algebranation.com

Why Use It?

Aligned specifically with **Florida Algebra 1 EOC standards**.

Includes **video tutorials, practice questions, and online study groups**.

Features **expert-led lessons** for step-by-step learning.

Khan Academy – Free Algebra 1 Course

https://www.khanacademy.org/math/algebra

Why Use It?

Interactive video tutorials on every Algebra 1 topic.

Instant feedback on practice problems.

Covers **linear equations, quadratics, functions, polynomials, and more**.

IXL – Algebra 1 Skill Practice

https://www.ixl.com/math/algebra-1

Why Use It?

AI-based personalized learning with **real-time progress tracking**.

Thousands of **practice questions with step-by-step explanations**.

Covers **function graphing, equations, probability, and more**.

Mathway – Algebra Problem Solver

https://www.mathway.com

Why Use It?

Allows students to **input algebra problems and receive instant step-by-step solutions**.

Useful for checking answers and understanding **how to solve different types of problems**.

Covers **basic algebra, polynomials, radical expressions, and quadratic equations**.

Desmos – Online Graphing Calculator

https://www.desmos.com

Why Use It?

Free online graphing calculator to visualize linear and quadratic equations.

Helps with **function transformations, inequalities, and real-world problem modeling**.

User-friendly and perfect for **practicing graph-based questions** in **Session 2 of the EOC Exam**.

Varsity Tutors – Free Algebra 1 Flashcards & Quizzes

https://www.varsitytutors.com/algebra_1-flashcards

Why Use It?

Free **Algebra 1 flashcards** for quick review of key formulas.

Includes **interactive quizzes** to test knowledge.

Great for **memorizing properties of numbers, exponents, and functions**.

OpenStax – Free Algebra 1 Textbook

https://openstax.org/details/books/algebra

Why Use It?

Free **full-length Algebra 1 textbook** with in-depth explanations.

Covers **every major topic in the Florida EOC exam.**

Includes **practice exercises** and **real-world applications**.

Recommended Academic Materials

Best Algebra 1 Textbooks & Study Guides

Title	Author/Publisher	Why It's Recommended?
Algebra 1: Common Core Edition	McGraw Hill	Covers all Florida Algebra 1 EOC topics with real-world applications.
Barron's Algebra 1 EOC Study Guide	Barron's Test Prep	Focuses on EOC-style practice questions and test-taking strategies.
Princeton Review Algebra 1 Workbook	The Princeton Review	Great for targeted practice, concept explanations, and review quizzes.
Big Ideas Math: Algebra 1	Ron Larson & Laurie Boswell	Step-by-step problem-solving strategies with engaging exercises.
Algebra 1 for Dummies	Mary Jane Sterling	Best for students who need **clear, simple explanations** of algebra concepts.

Florida Algebra 1 EOC Practice Tests & Workbooks

Florida Algebra 1 EOC Success Strategies Workbook

Why Use It?

✔ EOC-style practice questions with detailed answer explanations.

✔ Strategies specifically for **Florida's testing format**.

Florida Algebra 1 EOC Test Prep & Practice Questions (by Test Prep Books)

Why Use It?

✔ Full-length practice tests that **simulate the real EOC exam**.

✔ Great for **self-assessment and last-minute preparation**.

SAT & ACT Math Prep Books (For Additional Practice)

- **SAT Math Prep by The College Board**

- **ACT Math Prep by Kaplan**

- **SAT/ACT Math Flashcards by Barron's**

Why Use Them?

✔ Many **Algebra 1 topics** appear on the **SAT/ACT**, so these books help build **long-term math skills**.

✔ Great for **additional problem-solving techniques** beyond the EOC exam.

Final Tips for Using These Resources Effectively

How to Use These Resources for Maximum Benefit?

✔ **Use Online Tools for Quick Concept Review** – Watch **Khan Academy** and **Algebra Nation** videos to strengthen weak areas.

✔ **Take EOC-Style Practice Tests** – Use **FLDOE, Varsity Tutors, and Barron's Study Guide** for **realistic practice**.

✔ **Practice Graphing Problems with Desmos** – Many Algebra 1 EOC questions involve **graphing linear and quadratic functions**.

✔ **Check Answers with Mathway or IXL** – Use these tools to **verify solutions and identify mistakes**.

✔ **Review Key Concepts with Flashcards** – Use **Varsity Tutors' flashcards** to memorize formulas and properties.

Final Words

With **the right resources and study strategies**, you can **boost your confidence and excel on the Florida Algebra 1 EOC Exam!**

Use this guide, practice consistently, and take advantage of these recommended tools to strengthen your algebra skills.

You're on your way to Algebra 1 success!

FINAL WORDS

YOUR JOURNEY TO SUCCESS

Congratulations on taking the first step toward mastering **Algebra 1** and preparing for the **Florida Algebra 1 End-of-Course (EOC) Exam**! You've equipped yourself with the knowledge, strategies, and resources needed to succeed. Now, as you approach the final stretch, remember this:

You Are Capable of Greatness!

Mathematics can be challenging, but every problem you solve **makes you stronger and more confident**. Don't let temporary struggles discourage you—every mistake is a lesson, and every challenge is an opportunity to grow.

5 Key Mindsets for Success

Believe in Yourself

Your mindset is your greatest tool. If you tell yourself, "I can do this," you're already halfway there. You've put in the effort, reviewed the material, and practiced—**you are ready!**

Trust the Process

You have **studied, practiced, and prepared** using this guide.
You understand the **test format, question types, and strategies**.
You have built a **strong mathematical foundation** that will help you beyond this test.

Trust in what you've learned. Your hard work will pay off!

Overcome Challenges with a Positive Attitude

Mistakes don't define you—they guide you! If a problem seems difficult, take a deep breath and break it down step by step. **Stay calm and focused.** You've already conquered many math problems before, and this is just another challenge to solve.

Manage Your Time and Stay Confident

During the exam:

* Stay calm and focused.

* **Skip tough questions** and come back to them later.

* **Use elimination strategies** to narrow down multiple-choice options.

* **Review your answers** if you have extra time.

Confidence comes from preparation, and you have prepared well!

Look Beyond the Test — This is Just the Beginning!

The skills you are learning now will help you in:

✔ **Higher-level math courses (Geometry, Algebra 2, Pre-Calculus, Calculus).**

✔ **College entrance exams (SAT, ACT).**

✔ **Real-world problem-solving in careers like engineering, business, science, and technology.**

Mathematics is everywhere, and mastering Algebra 1 will open new doors for your future!

Words of Encouragement from Successful Students

Emma (Passed Algebra 1 EOC with a Level 4 Score)

"I struggled with algebra at first, but I didn't give up. Practicing every day helped me feel confident, and using test-taking strategies really made a difference on exam day!"

Liam (Improved from Level 2 to Level 3 on Retake)

"I failed my first attempt, but instead of giving up, I focused on my weak areas and practiced consistently. I retook the test and passed! If I can do it, so can you!"

Sophia (Scored Level 5 — Mastery Level)

"Algebra seemed intimidating, but once I started understanding patterns and solving problems step by step, I found it fun! This exam is just a stepping stone to bigger achievements!"

Exam Day Reminder — Stay Positive & Do Your Best!

✔ **Take deep breaths** before starting.

✔ **Stay confident**—you've got this!

✔ **Keep a steady pace**—don't rush, but don't get stuck for too long.

✔ **Check your answers**—small mistakes can make a big difference.

✔ **Leave the exam feeling proud**—you worked hard, and that's what matters!

You Are Ready! Go Crush This Exam!

Remember: Success isn't just about intelligence—it's about effort, perseverance, and believing in yourself. No matter what, you are already a winner for preparing, learning, and growing.

Best of luck! You can do this!

EXPLORE OUR RANGE OF STUDY GUIDES

At Test Treasure Publication, we understand that academic success requires more than just raw intelligence or tireless effort—it requires targeted preparation. That's why we offer an extensive range of study guides, meticulously designed to help you excel in various exams across the USA.

Our Offerings

- **Medical Exams:** Conquer the MCAT, USMLE, and more with our comprehensive study guides, complete with practice questions and diagnostic tests.

- **Law Exams:** Get a leg up on the LSAT and bar exams with our tailored resources, offering theoretical insights and practical exercises.

- **Business and Management Tests:** Ace the GMAT and other business exams with our incisive guides, equipped with real-world examples and scenarios.

- **Engineering & Technical Exams:** Prep for the FE, PE, and other technical exams with our specialized guides, which delve into both fundamentals and complexities.

- **High School Exams:** Be it the SAT, ACT, or AP tests, our high school range is designed to give you a competitive edge.

- **State-Specific Exams:** Tailored resources to help you with exams unique to specific states, whether it's teacher qualification exams or state civil service exams.

Why Choose Test Treasure Publication?

- **Comprehensive Coverage:** Each guide covers all essential topics in detail.

- **Quality Material:** Crafted by experts in each field.

- **Interactive Tools:** Flashcards, online quizzes, and downloadable resources to complement your study.

- **Customizable Learning:** Personalize your prep journey by focusing on areas where you need the most help.

- **Community Support:** Access to online forums where you can discuss concerns, seek guidance, and share success stories.

Contact Us

For inquiries about our study guides, or to provide feedback, please email us at support@testtreasure.com.

Order Now

Ready to elevate your preparation to the next level? Visit our website www.testtreasure.com to browse our complete range of study guides and make your purchase.

Made in United States
Orlando, FL
31 March 2025

60008237R00153